Table of Contents

Unit I—Electrical Basics

Unit II—Electrical Distribution for Industry

Unit III–Electric Motor Basics

Unit IV–Electrical Motor Control

Unit V–Electronic Control Basics

Laboratory Manual

Electrical Motor Control Systems

Electronic and Digital Controls Fundamentals and Applications

by

Dale R. Patrick
Stephen W. Fardo

Publisher
The Goodheart-Willcox Company, Inc.
Tinley Park, Illinois

Unit VI—Industrial Electronic Control Systems

Unit VII—Digital Control, PLCs, Robotics, Troubleshooting, and Preventive Maintenance

Introduction

This laboratory manual is designed for use with the Electrical Motor Control Systems textbook. The activities in this manual are closely tied to the textbook. The activities have been divided into seven units in parallel with the text.

Unit I—Electrical Basics

Unit II—Electrical Distribution for Industry

Unit III—Electric Motor Basics

Unit IV—Electrical Motor Control

Unit V—Electronic Control Basics

Unit VI—Industrial Electronic Control Systems

Unit VII—Digital Control, PLCs, Robotics, Troubleshooting, and Preventive Maintenance

Each laboratory reinforces the important concepts in the corresponding unit from the textbook.

Each laboratory begins with your objective and a list of all the equipment you will need to complete the activity. Following the list of equipment, are step-by-step instructions that take you through the laboratory. In these laboratories, you are expected to set up electrical circuits, test motors, collect data, and answer questions. Charts are provided in many of the activities for your data.

Working with electricity and motors can be dangerous. Always use safe laboratory practices. If you have any questions about safe laboratory practices, discuss them with your instructor.

Hands-on learning is very important. These laboratory activities are designed to increase your understanding of electrical motor control. Take your time while doing these activities. Your knowledge and skills will be enriched.

Dale R. Patrick

Stephen W. Fardo

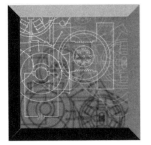

Activity 1–Electrical Conductors

Name _____ Date _____ Score _____

Objectives

In this activity, you will measure the diameter of several electrical conductors using a wire gage.

Equipment and Materials

- Wire gage
- Electrical conductors of various sizes

Procedure

1. Obtain an wire gage (see Figure 1-1) and several conductors from your teacher. The wire gage will give conductor readings in AWG (American Wire Gage).

2. Measure the diameter of each conductor with the gage.

3. Record the values on a sheet of paper.

4. Below the AWG value, record whether the conductor is copper or aluminum.

5. If an insulation type is marked on the wire, record the insulation type also.

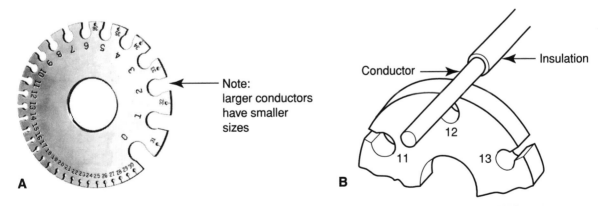

Figure 1-1. A—American Wire Gage is used to measure conductor size. B—Using an AWG.

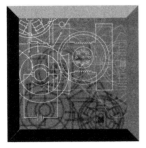

Activity 2–Electrical Tool Identification

Name _____ Date _____ Score _____

Objectives

In this activity, you will learn to correctly identify the types of tools used by electrical technicians. It is important to know the correct names of the tools used by a technician.

Equipment and Materials

- Electrical tools used in your lab or shop

Procedure

1. Study the illustrations of electrical tools in Chapter 2 of your textbook.

2. Learn the *correct* names of these tools.

3. Learn the uses of the tools as you study them.

4. Several common tools are shown in Figure 2-1. Fill in the correct names of the tools shown in Figure 2-1.

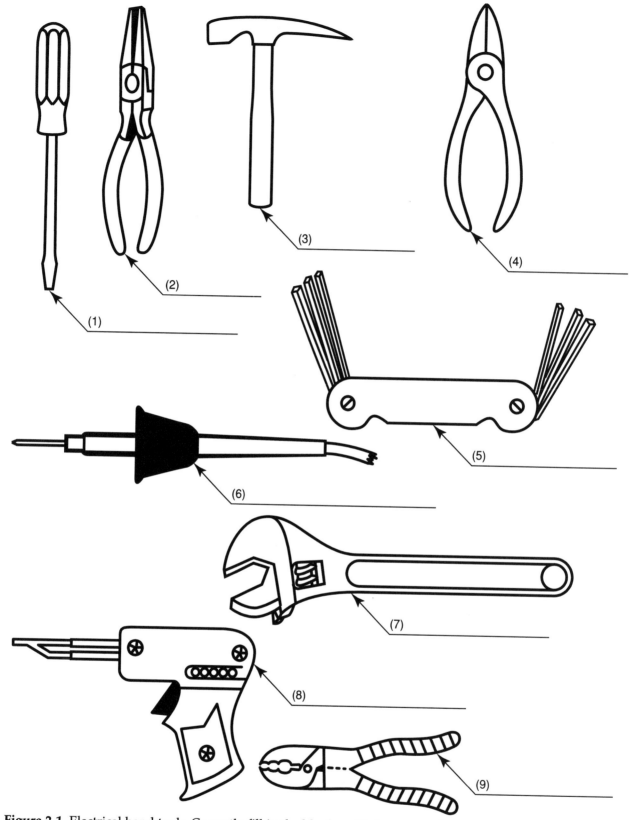

Figure 2-1. Electrical hand tools. Correctly fill in the blanks with the name of each tool.

5. Your instructor can now give you a test to see how many of the electrical tools used in your laboratory or shop you can identify.

Activity 3–Soldering and Terminal Connections

Name _____ Date _____ Score _____

Objectives

In this activity, you will demonstrate your skill in three important exercises: (1) soldering, (2) desoldering, and (3) terminal connection. An understanding of the proper soldering and terminal connection techniques is needed by technicians who design, service, or repair industrial equipment, including robots.

Equipment and Materials

- Soldering iron—25 to 50 W
- Desoldering tool
- Solder
- No. 22 insulated wire—3" length
- Printed circuit board
- Component-mounting strip
- Resistor (any value)
- Wire stripper
- Side-cutting pliers
- Needle-nose pliers
- "Solderless" connector and wire
- Crimping tool

Safety

Exercise caution while soldering. Use protective eyewear and clothing during soldering operations.

Procedure

Section A. Soldering Iron Preparation and Component Mounting

1. Obtain a soldering iron and some solder.

2. Allow the soldering iron to heat to its operating temperature.

3. Apply a small amount of solder to tin the tip of the iron.

4. Obtain a component-mounting strip and resistor from your instructor.

5. Using the proper soldering techniques found in Chapter 2 of your textbook, solder the resistor onto the component-mounting strip.

6. Show the component-mounting strip to your instructor for approval.

   Instructor's Approval: _____

Section B. Printed Circuit Board Soldering.

1. Obtain a printed circuit board, a 3" length of No. 22 insulated wire, and a pair of wire strippers.

2. Place the wire near the holes in the printed circuit board where you intend to solder it to determine the length of wire required.

3. Strip 3/8" to 1/2" of insulation from one end of the wire with the wire stripper.

4. Cut off any excess wire and strip the insulation off the other end of the wire. The wire should lie flat on the printed circuit board when soldered.

5. Using the proper soldering techniques, solder both ends of the wire to the printed circuit board terminals.

6. Show the printed circuit board to your instructor for approval.

   Instructor's Approval: _____

Section C. Printed Circuit Board Desoldering

1. Obtain a desoldering tool.

2. Allow the desoldering tool to heat to its operating temperature.

3. Using the proper desoldering techniques discussed in Chapter 2 of your textbook, remove the piece of wire from the printed circuit board without damaging the conductive strips on the board.

4. Show the printed circuit board to your instructor for approval.

   Instructor's Approval: _____

Section D. Terminal Connection Installation

1. Obtain a "solderless" terminal connector crimping tool and a length of insulated wire from your instructor.

2. Strip enough insulation from the wire so that the terminal connector fits properly.

3. Use the crimping tool to fasten the terminal connector to the wire.

4. Show the terminal connector and wire to your instructor for approval.

   Instructor's Approval: _____

5. Return all unused materials.

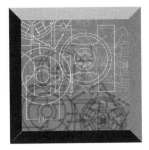

Activity 4–Basic Electrical Symbols

Name _____ Date _____ Score _____

Objectives

Working with industrial control systems requires the ability to read and recognize electrical schematic symbols. Electrical schematics are used to maintain, adjust, and repair robotic systems and controllers. Many of these symbols will be used in the activities that follow.

Coil (air core)		DC generator	
Coil (iron core)		AC motor	
Single cell		Wires (connected)	or
Battery		Wires (crossing)	or
Galvanometer	or	Push-button switch (normally open)-NO	
Voltmeter		Push-button switch (normally closed)-NC	
Ammeter		Switch single-pole, single-throw (SPST)	
Ohmmeter		Switch double-pole, single-throw (DPST)	
Wattmeter		Switch single-pole, double-throw (SPDT)	
AC source		Transformer	Primary Secondary
Neon lamp		Fuse	
Incandescent lamp		Variable resistor (potentiometer)	
Ground		Fixed resistor	
Antenna		Relay	Coil — Normally closed contact / Normally open contact
Fixed capacitor		Transistor (PNP)	B=Base C=Collector E=Emitter
Variable capacitor			
Diode (rectifier)		Transistor (NPN)	

Figure 4-1. Electrical schematic symbols.

15

In this activity, you should study and learn to recognize the basic symbols shown in Figure 4-1. You can demonstrate your understanding by completing the self-test on electrical symbols.

Analysis

Without looking at the symbols in Figure 4-1, identify each of the symbols that follow. Place the correct response in the space provided.

1. —⎺w⎺— _____

2. —(A)— _____

3. —(OHM)— _____

4. —∕∘— _____

5. ═⊖ _____

6. —▸⊦⊦— _____

7. —⊥— _____

8. —⊥— _____

9. —ⱳ↑— _____

10. —⊣⊦⊦— _____

11. —(V)— _____

12. —⊣⊦— _____

Activity 5–Electrical Components, Equipment, and Symbols

Name _____ Date _____ Score _____

Objectives

Basic to the study of any technical subject is the understanding of the language and symbols used. The study of control technology is dependent on the graphic language of electrical diagrams. Various components, equipment, and symbols are used. You should become familiar with the graphic symbols used in electrical diagrams and learn to recognize the symbols used for common electrical devices.

Equipment and Materials

- Multimeter
- Soldering iron—25 to 50 W
- SPST switch
- Potentiometer (any value)
- Connecting wires for circuit board
- Battery—6 volt
- Lamp with socket—6 volt

Safety

Exercise caution while soldering. Use protective eyewear and clothing during soldering operations.

Procedure

1. The wires you will use in this activity are paths for the movement of electrons. Wires can be connected to each other at almost any angle and can cross each other without being connected. The graphic symbols representing these wires, called conductors, are illustrated in Figure 5-1.

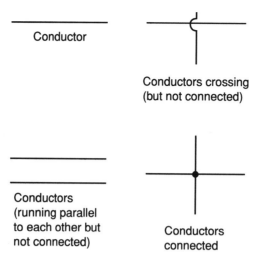

Conductor

Conductors crossing
(but not connected)

Conductors
(running parallel
to each other but
not connected)

Conductors
connected

Figure 5-1. Different schematic symbols for conductors.

2. In the illustration of Figure 5-2, how many times do conductors cross each other without connecting and how many times are conductors connected to each other?

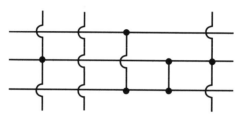

Figure 5-2. Various conductors in a schematic.

Conductors cross each other _____ times.

Conductors are connected to each other _____ times.

3. A single-pole, single-throw (SPST) switch is a device used to allow current to flow when closed, or on, and interrupt the flow of current when open, or off. A switch is used to turn the light in a room on and off. The symbols reprenting the SPST switch in its two conditions, on and off, are illustrated in Figure 5-3.

SPST "on"
or "closed"

SPST "off"
or "open"

Figure 5-3. Schematic symbols for a single-pole, single-throw switch.

4. Fixed resistors are used frequently to control current. The symbol for a fixed resistor is illustrated in Figure 5-4.

Figure 5-4. Fixed resistor.

Activity 5—Electrical Components, Equipment, and Symbols

5. A light bulb or lamp is a common component used in flashlights and other devices to produce light. The symbol of an incandescent lamp is shown in Figure 5-5.

Figure 5-5. Incandescent lamp.

6. A potentiometer, or pot, is a resistive component that can be adjusted to control current. The illustrations of Figure 5-6 represent the schematic symbols for a potentiometer. Notice that the center connection of the potentiometer is designated by an arrow. It represents the adjustable portion of the device.

Potentiometer   Potentiometer

Figure 5-6. Schematic symbols for a potentiometer.

7. The 6-V battery is a chemical source of electrical energy that causes current to flow through conductors, resistors, and other components. The symbol of Figure 5-7 is used for a battery. Note that one of the connectors of the battery is labeled with a + sign and the other is labeled with a – sign. It is always important to observe how a battery is connected to a circuit.

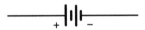

Figure 5-7. Battery.

8. Use your circuit board or trainer unit to set up the circuit shown in Figure 5-8. Two types of circuit boards that can be used for your experiment are shown in Figure 5-9. You should study the layout of the board you are using. Be sure that you connect this circuit and the others you will construct properly. You should become very familiar with the uses of a circuit board. For most circuit boards, it is necessary to cut wire of the proper diameter to various sizes and strip about 1/4 in. of insulation from both ends. The wires are then used to make circuit connections. The switch and potentiometer should have wires of about 2 in. long soldered to their terminals for connecting to the circuit board. They will be used for other experiments.

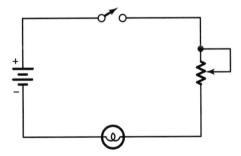

Figure 5-8. Series circuit with potentiometer.

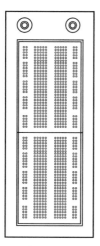 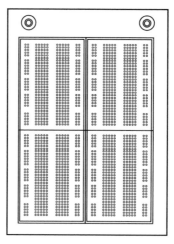

Figure 5-9. Circuit trainer board.

9. Complete the circuit connection using your board. Close the SPST switch and adjust the potentiometer from its maximum counterclockwise position to its maximum clockwise position. What happens to the lamp? If the circuit is constructed properly, the lamp should get brighter and then dimmer as the potentiometer is adjusted.

10. Adjust the potentiometer until the lamp is at its brightest and open the SPST switch. What happened to the lamp?

Analysis

1. Draw the symbols for the indicated components in the space that follows.
   Fixed Resistor

   Potentiometer

   Open SPST switch

2. Draw the symbols for conductors in the space below.

Conductors crossing (not connected)

Conductors connected

3. Draw the symbol for a battery, placing the positive sign and the negative sign at the proper sides of the symbol.

4. Draw the symbol for a lamp.

Activity 6–Resistor Color Code

Name _____ Date _____ Score _____

Objectives

Resistors are one of the most often used electrical control components. In this activity, you will review the resistor color code and show your knowledge by identifying the values of several resistors on a self-test. Study the resistor color codes shown below, then complete the self-test.

Resistor Color Code

Color	1st Significant Figure	2nd Significant Figure	Multiplier	Tolerance
Black	0	0	1	
Brown	1	1	10	
Red	2	2	100	
Orange	3	3	1000	
Yellow	4	4	10,000	
Green	5	5	100,000	
Blue	6	6	1,000,000	
Violet	7	7	10,000,000	
Gray	8	8		
White	9	9		
Gold*	—	—	0.1	5%
Silver*	—	—	0.01	10%
No color	—	—		20%

1st significant figure

2nd significant figure

Multiplier

Tolerance

a b c d

Resistor

* When resistors have a value of less than 10 ohms, the third color band is a decimal multiplier. The two colors used are: gold = ×0.1 and silver = ×0.01.

Resistor Color Code–Self Test

For each of the following color codes, identify the resistance value and tolerance of the resistor. Place your answer in the proper blank space.

1. (a) red
   (b) red
   (c) blue

   1. Answer:_____ ohms, _____ tolerance.

2. (a) black
   (b) brown
   (c) brown
   (d) gold

   2. Answer:_____ ohms, _____ tolerance.

3. (a) gray
   (b) red
   (c) black
   (d) silver

   3. Answer:_____ ohms, _____ tolerance.

4. (a) white
   (b) brown
   (c) orange
   (d) gold

   4. Answer:_____ ohms, _____ tolerance.

5. (a) yellow
   (b) violet
   (c) orange

   5. Answer:_____ ohms, _____ tolerance.

6. (a) brown
   (b) green
   (c) orange

   6. Answer:_____ ohms, _____ tolerance.

7. (a) blue
   (b) gray
   (c) black
   (d) gold

   7. Answer:_____ ohms, _____ tolerance.

8. (a) green
   (b) blue
   (c) silver
   (d) gold

   8. Answer:_____ ohms, _____ tolerance.

9. (a) brown
   (b) black
   (c) black

   9. Answer:_____ ohms, _____ tolerance.

10. (a) orange
    (b) white
    (c) black
    (d) gold

    10. Answer:_____ ohms, _____ tolerance.

11. (a) red
    (b) violet
    (c) gold
    (d) gold

11.   Answer:_____ ohms, _____ tolerance.

12. (a) brown
    (b) brown
    (c) black
    (d) gold

12.   Answer:_____ ohms, _____ tolerance.

13. (a) violet
    (b) green
    (c) orange
    (d) gold

13.   Answer:_____ ohms, _____ tolerance.

14. (a) yellow
    (b) violet
    (c) green

14.   Answer:_____ ohms, _____ tolerance.

15. (a) green
    (b) blue
    (c) orange

15.   Answer:_____ ohms, _____ tolerance.

16. (a) blue
    (b) gray
    (c) orange

16.   Answer:_____ ohms, _____ tolerance.

17. (a) yellow
    (b) orange
    (c) brown
    (d) gold

17.   Answer:_____ ohms, _____ tolerance.

18. (a) orange
    (b) orange
    (c) red
    (d) gold

18.   Answer:_____ ohms, _____ tolerance.

19. (a) brown
    (b) black
    (c) green

19.   Answer:_____ ohms, _____ tolerance.

20. (a) red
    (b) red
    (c) red

20.   Answer:_____ ohms, _____ tolerance.

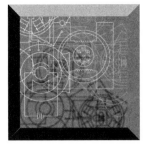

Activity 7–Measuring Resistance

Name _____ Date _____ Score _____

Objectives

Resistance is the term used to describe the opposition encountered by electrical current. Resistance is measured in ohms (Ω).

In the previous activity, you learned how to determine the value of a resistor by observing its color bands. However, the resistance of many components cannot be determined by observation and therefore must be measured. In this activity, you will learn how to measure resistance with the ohmmeter portion of a multimeter. If you are using a digital meter, use this activity to learn to read an analog scale to measure resistance.

Equipment and Materials

- Multimeter
- Resistors—10 Ω, 15 Ω, 220 Ω, 470 Ω, 1 kΩ, 5.1 kΩ, 68 kΩ, 100 kΩ, 220 kΩ, 1 MΩ
- Potentiometer—200 Ω

Procedure

1. The multimeter is the most used meter for basic electronic work. It can be used as an ammeter, a voltmeter, or an ohmmeter. Simply adjust the function selection switch for the desired function. Figure 7-1 shows the controls, including the function-select switch, of a common analog multimeter.

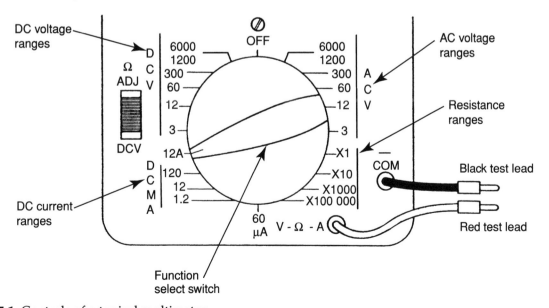

Figure 7-1. Controls of a typical multimeter.

You will notice that the function-select switch is in the center of some circular divisions and acts as a dial. You will also notice that a portion of the circular divisions is designated as the ohms function. Further, notice that the area of the divisions located within the ohms-function space is divided into four positions: ×1, ×100, ×1000, and ×100,000. The meter that you are using may be slightly different from the one illustrated. Nevertheless, it will exhibit the same basic characteristics as the multimeter shown. Adjust the function-select switch on your meter to the ohms function. List below the different positions within the ohms-function area on your meter.

_____

_____

2. Now that you have adjusted the meter to the ohms function, connect the test leads to the meter. Notice that the test leads used with the meter are (ordinarily) red and black. The colors are used in order to help you distinguish between the positive and negative polarities of the meter. For our purposes, red indicates the positive polarity (+), and black indicates the negative (–) polarity. Insert the red test lead in the hole or jack marked with a (+) and the black test lead in the hole or jack marked with a (–). (*Note*: The meter you are using may be different from the one illustrated in Figure 7-1. If this is so, plug the red and black test leads into the appropriate jacks and proceed.) Touch the test leads together and describe what happens to the needle located on the scale of the meter.

_____

_____

3. You have now adjusted the function selection switch to the ohms function of the meter, inserted the proper test leads into the proper jacks, and seen that when the test leads are touched together, or shorted, the needle of the meter moves from the extreme left-hand side of the meter scale to the extreme right-hand side. If you are using a digital meter, the display will indicate 0. When the leads are not touched together, the digit 1 will appear on the left display in most cases.

4. You should now become familiar with the scale of the meter. Figure 7-2 is a scale of a multimeter. If you are using a digital meter, use this section as a review of analog scales.

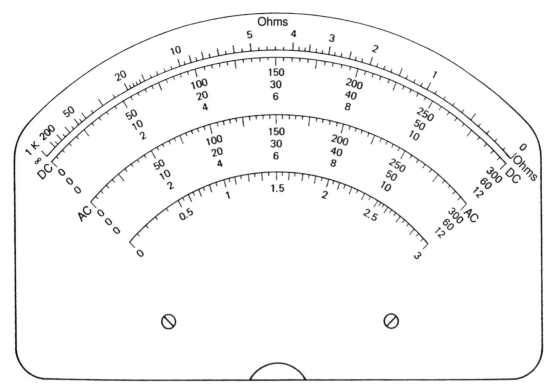

Figure 7-2. Multimeter scale.

You will notice that the top scale (from 0 to ∞, labeled *ohms*) is used for measuring ohms only. On most meters, the top scale is designated as the ohms scale. Record the location and color of the ohms scale used on your meter.

Location	Color

5. To measure any resistance in ohms, you must select the proper position or range. On the meter illustrated in Figure 7-1, there are four ranges: ×1, ×10, ×1000, and ×100,000. The ohmmeter must be properly *zeroed* before attempting to accurately measure resistance. To zero the ohmmeter, short the two test leads together. This should cause the needle to move from infinity (∞) on the left to zero (0) on the right. Infinity represents a very *high* resistance, whereas zero represents a very *low* resistance. If the needle does not reach zero or if it goes past zero when the test leads are shorted, then the control marked *ohms adjust* must be used to adjust the needle to rest on zero when the test leads are touched together. There will be a similar adjustment on your meter, if yours is different from the one illustrated.

   Adjust the meter to each of its ranges: ×1, ×10, ×1000, and ×100,000 and zero the meter for each of these ranges. The ohmmeter must be zeroed prior to each resistance measurement and after each range change. Otherwise your measurements will be incorrect.

6. Notice that the ohms scale on the meter is *nonlinear*. The divisions on the right side of the scale are farther apart than those on the left side. This provides a more accurate measurement of resistance when the needle of the meter deflects and stops somewhere between the center of the ohms scale and zero. Choosing the proper range adjustment will control where the needle deflects. Adjust the function-select switch to the ×1 range of the ohms function. Zero the meter, and then connect it across a 10-Ω resistor. Record the precise resistance (in ohms) indicated by the needle on the ohms scale in Figure 7-3. The range selected for this operation was ×1, which means that the number to which the needle points will be multiplied by 1.

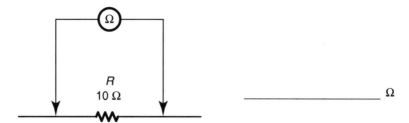

Figure 7-3.

7. Adjust the function-select switch to the ×100,000 range. Zero the meter and then connect it to a 100-kΩ resistor, as illustrated in Figure 7-4. Record the resistance (in ohms) shown by the needle on the ohms scale. The range selected for this measurement was ×100,000, which means that the number to which the needle points will be multiplied by 100,000.

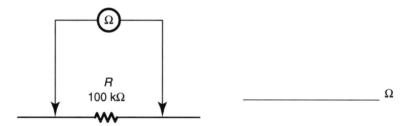

Figure 7-4.

8. Measure and record the values of the resistors indicated in Figure 7-5. Remember to choose a meter range that will cause the needle to deflect somewhere between the center of the scale and zero. Always zero the meter when changing ranges, and always multiply the number indicated on the scale by the multiplier of the chosen range: ×1, ×10, ×1000, and ×100,000. Never measure the resistance of a component until it has been disconnected.

Color-coded value	Measured value in ohms
15 Ω	
100 Ω	
220 Ω	
470 Ω	
1 kΩ	
5.1 kΩ	
68 kΩ	
220 kΩ	
1 MΩ	

Figure 7-5.

9. Using the proper procedure for measuring resistance, measure and record the precise resistance of the 200-Ω potentiometer illustrated in Figure 7-6.

Activity 7—Measuring Resistance

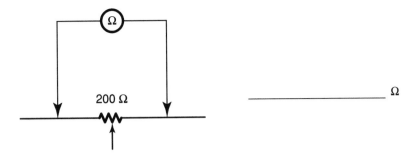

Figure 7-6.

10. Adjust the control of the potentiometer while the ohmmeter is connected. Describe how this action affects the measured resistance of the potentiometer. _____

_____

11. Alter the connections of the potentiometer as illustrated in Figure 7-7.

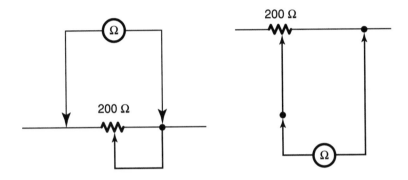

Figure 7-7.

12. Adjust the potentiometer both clockwise and counterclockwise and describe how this action affects its measured resistance.

Analysis

1. Why is the ohms scale of the multimeter considered to be nonlinear? _____

_____

2. Where on the ohms scale are the most accurate measurements found? _____

_____

3. What is meant by the ×1000 range on the ohmmeter? _____

_____

4. What is meant by *zeroing* the ohmmeter? _____

_____

5. Why is it necessary to zero the ohmmeter? _____

_____

6. If the range of the ohmmeter was set to ×100,000 and the needle pointed to 0.6 on the ohms scale, what would be the value of the resistance being measured? _____

_____

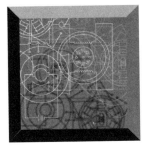

Activity 8–Measuring Voltage

Name _____ Date _____ Score _____

Objectives

Voltage is the force that causes electrical current to flow through conductors or paths. This force, sometimes called electromotive force (EMF), is measured in volts. It is essential that anyone involved in electronic work be able to use the necessary equipment to measure voltage accurately. In this activity you will:

1. Learn to measure voltage with a voltmeter.

2. Construct basic electrical circuits.

If you are using a digital meter, use this activity to learn to read an analog scale to measure voltage.

Equipment and Materials

- Multimeter
- Resistors—220 Ω, 470 Ω, 1 kΩ
- Potentiometer—200 Ω
- Battery or power supply—6 volt
- Lamp with socket—6 volt
- Connecting wires

Procedure:

1. You learned in the previous activity that a multimeter can be used to measure resistance. Likewise, a multimeter can be used to measure voltage. You will also recall from the previous activity that the function-select switch can be adjusted to cause the meter to perform many measurement functions using several different ranges. When the function-select switch is adjusted to 3 V on the dc volts range, the meter is capable of measuring a *maximum* of 3 V. The same is true for the remaining ranges within the dc volts function. The numerical value of the chosen range indicates the maximum value of voltage that can be measured on the range. The meter you are using may differ somewhat from the multimeter that is illustrated in Figure 8-1. Nevertheless, it will exhibit the same basic characteristics as the one described.

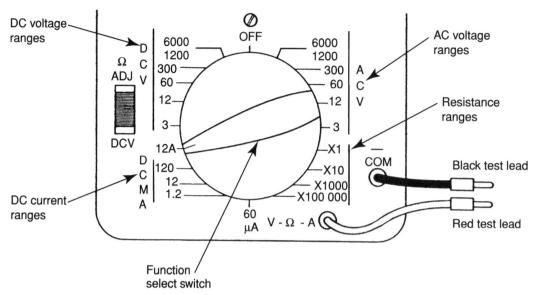

Figure 8-1. Sketch of multimeter settings.

2. Adjust the function-select switch on your meter to the lowest range of the dc volts function. List the different ranges within the dc volts function on your meter.

_____, _____, _____, _____, _____, _____.

3. Connect the red and black test leads to the meter by inserting the appropriate ends into the proper jacks on the face of the meter. The red test lead should be inserted into the jack labeled + and the black test lead should be inserted into the jack labeled – COM. (*Note:* The meter you are using may be different. If this is the case, plug the red and black test leads into the appropriate jacks and proceed.)

4. You now have the multimeter properly equipped and adjusted to measure dc volts. You should now become familiar with that portion of the scale of the meter used for measuring dc volts. Part A of Figure 8-2 shows the scale of a very commonly used multimeter. Notice that the scale immediately under the ohms scale is the dc volts scale. You will also notice that there are three dc volts scales: 0–12 V, 0–60 V, and 0–300 V. All dc voltages are measured using one of these scales. The question becomes, what scales are to be used with what meter ranges? Notice that each of the dc voltage ranges corresponds to a number on the right side of the meter scale or a multiple or divisor of that number.

5. You will notice that when you choose the 12-V, 60-V, or 300-V range, you read the scale directly. On these ranges, the number to which the needle points is the actual value of the voltage being measured. When you select the 3-V range, the number to which the needle points must be divided by 100. For example, if the needle of the meter points to the number 50 while the meter is adjusted to the 3-V range, the measured voltage is 0.5 V. (Note: The meter you are using may differ slightly from the one just described. Generally the meter you are using will have several scales. Some of these scales can be read directly, whereas others will require the use of a multiplier or divisor. If your meter is different, identify the scales which can be read directly and the ones which must be used with multipliers or divisors.)

6. Complete Part B of Figure 8-2 by indicating the proper voltage scale and voltage value. Also show the multiplier or divisor for the voltage range for the meter scale shown. This will give you practice using a meter scale.

Activity 8—Measuring Voltage

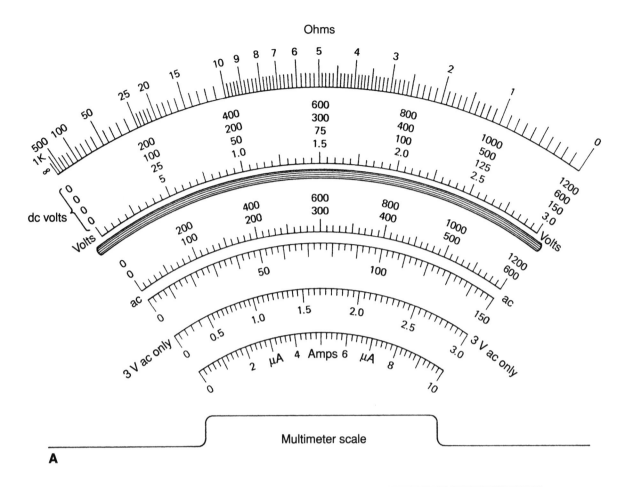

A

Multimeter scale

Multimeter range	Proper scale to read	Number needle points to	Voltage value	Multiplier	Divisor
Example: 3 V	0 to 3	1.5	1.5	None	None
3 V		2.0			
15 V		100			
60 V		300			
150 V		75			
600 V		200			
1200 V		800			
15 V		125			
60 V		350			
150 V		125			
60 V		125			
15 V		75			

B

Figure 8-2. Multimeter scale and range settings.

7. You are now almost ready to measure dc voltage. Before actually doing so you should become familiar with some facts about choosing the dc voltage range. You will recall from step 1 that the numerical value of the chosen range indicates the *maximum* value of voltage that can be measured on that range. When the range selected is 12 V, the maximum voltage that the meter can measure is 12 V. Any voltage above 12 V could damage the meter on this range. This is true for most voltage scales on most multimeters. If you are trying to measure a voltage that is totally unknown (no indications as to its approximate value), you should start the measuring procedure by choosing the *highest* range on your meter and slowly adjusting the range downward until a voltage reading is indicated on the upper or right-hand half of the meter scale. Complete Figure 8-3 by showing the proper meter range to be selected, the proper meter scale to be read, and the proper multiplier and divisor to be used with each voltage to be measured. Use the scale of your laboratory meter and complete the figure.

Voltage to be measured	Range to select	Scale to read	Multiplier	Divisor
Example: 2.0 V	3V	0 to 3.0	None	None
3.5 V				
0.5 V				
6 V				
14 V				
17 V				
30 V				
65 V				
100 V				
250 V				
1100 V				
Unknown				

Figure 8-3. Complete the chart.

8. You also need to consider meter *polarity* when measuring dc voltage. Correct matching of meter polarity to the voltage polarity is vitally important. If proper care is not exercised, the meter will deflect backwards, causing damage to its internal components. Meter polarity is simple to determine. The positive (+) red test is connected to the positive side of the dc voltage to be measured. The negative (–) black test lead is connected to the negative side of the dc voltage to be measured.

9. Measure and record the actual voltage of a 6-V battery. The voltmeter is always connected *across* the voltage, or in parallel with the voltage to be measured.

   Actual voltage of battery = _____ V

10. Construct the circuit illustrated in Figure 8-4. Measure and record the voltage supplied by the battery and the voltage across the lamp.

Activity 8—Measuring Voltage

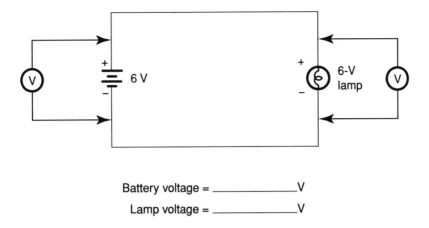

Battery voltage = _____V

Lamp voltage = _____V

Figure 8-4.

11. A certain amount of voltage is always necessary to cause current to flow. This voltage is called the *voltage drop* and is found across any component through which current flows. The polarity of this voltage drop is determined by the direction of the movement of electrons from negative to positive. Thus, if electrons were moving through a resistor in the direction shown in Figure 8-5, the left side of the resistor would be negative (–) and the right side would be positive (+). The black test lead of the meter would be connected to the left side of the resistor and the red test lead would be connected to the right side. This procedure would measure the voltage drop developed across this resistor. In the illustrations of Figure 8-6, indicate the proper meter polarity for measuring the voltage drop across each resistor.

Direction of electron movement

Figure 8-5.

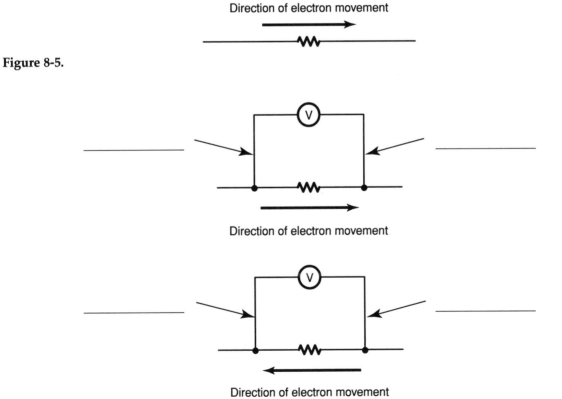

Figure 8-6.

12. Construct the circuit of Figure 8-7. Measure and record the battery voltage and the voltage drops across R_1 and R_2. Remember that electrons move from negative to positive.

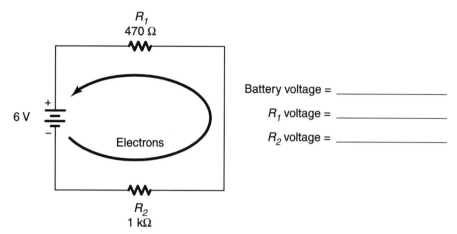

Battery voltage = _____

R_1 voltage = _____

R_2 voltage = _____

Figure 8-7.

13. Construct the circuit of Figure 8-8. Measure and record the voltage drops across R_1 and R_2.

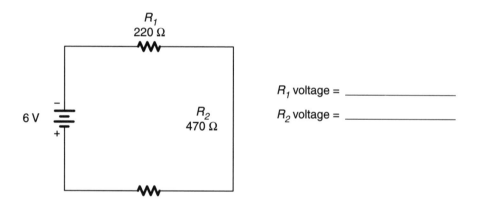

R_1 voltage = _____

R_2 voltage = _____

Figure 8-8.

14. Construct the circuit shown in Figure 8-9.

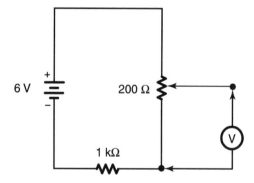

Figure 8-9.

15. Adjust the potentiometer both counterclockwise and clockwise and record the voltages.

Counterclockwise voltage = _____ V

Clockwise voltage = _____ V

Analysis

1. What is voltage? _____

   _____

2. If you were measuring an unknown voltage, what meter range would you choose? Why? ____

   _____

   _____

3. What determines the polarity of a voltage drop developed by a resistor through which electrons are moving? _____

   _____

4. What causes the voltage drop developed across any component through which electrons are moving? _____

   _____

5. When is it necessary to use a multiplier or divisor with a scale on a multimeter? _____

   _____

6. How is the proper range of a multimeter selected? _____

   _____

   _____

7. Could 4 V be measured on the 3-V range of a multimeter? Why? _____

   _____

8. Indicate the proper range of your meter for measuring the following dc voltages:

   10 V = _____ range

   2.8 V = _____ range

   6.0 V = _____ range

   100 V = _____ range

   65 V = _____ range

   40 V = _____ range

Activity 9–Measuring Current

Name _____ Date _____ Score _____

Objectives

Current is the movement of electrons from one location to another through conductors or paths. Current is measured in amperes (the number of electrons moving past a given point in a circuit per second), milliamperes (0.001 A), or microamperes (0.000001 A). It is essential that anyone involved in electronic work learn to measure current values accurately. In this activity, you will:

1. Learn to measure current with an ammeter.

2. Construct basic electrical currents.

3. Convert amperes to milliamperes or microamperes and vice versa.

If you are using a digital meter, use this activity to learn to read an analog scale to measure current.

Equipment and Materials

* Multimeter

* Resistors—220 Ω, 1 kΩ, 100 kΩ, 220 kΩ

* 6-V battery or power supply

* SPST switch

* Connecting wires

Procedure

1. You have learned from previous activities that a multimeter can be used to measure resistance and voltage. In addition to these electrical quantities, a multimeter can be used to measure current. Figures 9-1 and 9-2 show the controls and the scale of a common multimeter. The function selection switch controls the measurement function as well as the range of the meter. Figure 9-1 indicates that the function-select switch can be adjusted to one of five ranges within the dc function, namely, 12 A, 120 mA, 12 mA, 1.2 mA, and 60 μA. Thus, when the function-select switch is placed in the 120-mA position within the dc function, the meter is capable of measuring a maximum of 120 mA. Likewise, the other dc current ranges of the multimeter are similarly arranged. The numerical value of the chosen range shows the maximum value of current that can be measured on that range.

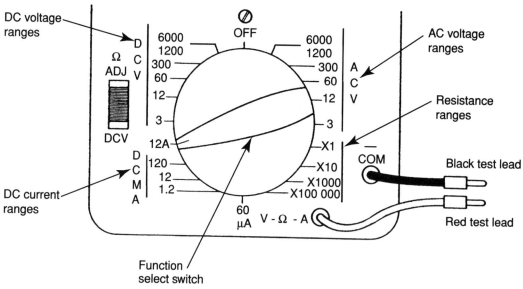

Figure 9-1. Controls of a typical multimeter.

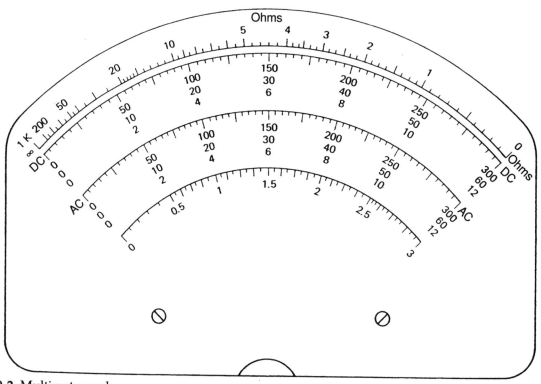

Figure 9-2. Multimeter scale.

2. The meter you are using may differ somewhat from the illustrated multimeter. However, it should exhibit the same basic characteristics as the one described. Adjust the function selection switch on your meter to the lowest range of the direct current function. List below the different ranges for measuring direct current on your meter.

_____, _____, _____, _____, _____, _____.

3. Plug the black and red test leads into the jacks labeled (–) and (+) respectively. (Note: If your meter is different from the one illustrated, plug the test leads into the appropriate jacks and proceed.)

Activity 9—Measuring Current

4. You are now ready to become familiar with the portion of the scale of the meter used for measuring direct current. Figure 9-2 illustrates the scale of a multimeter. Notice that there are three scales used for measuring dc. They are the same as those used to measure dc voltage. There are five dc ranges that can be used with this meter. All current measurements will be read on these scales, using the proper multiplier or divisor. Figure 9-3 shows the proper multiplier or divisor to be used with the dc scales, for any of the five ranges. (Note: Your meter may have different scales and multipliers from those in Figure 9-3. You should identify them and proceed.)

Multimeter current range	Multimeter scale	Multiplier	Divisor
60 µA	0-60	None	None
1.2 mA	0-12	—	10
12 mA	0-12	None	None
120 mA	0-12	10	—
12 A	0-12	None	None

Figure 9-3.

5. Observe that each range of dc has a specific multiplier or divisor or is read directly on the scale. For example, if the needle of the meter points to the number 6 on the dc scale, and the meter is adjusted to the 12-mA range, the measured current is 6 mA. If the needle points to 6 while the meter is adjusted to the 120-mA range, the measured current is 60 mA ($6 \times 10 = 60$, or $6 \div 0.1 = 60$).

6. Complete Figure 9-4 by determining the proper current value when the needle points to a specific number. Also, determine the correct multiplier and divisor to be used with the indicated current range. You should use the equivalent ranges, scales, multipliers, and divisors of the meter of Figures 9-1 and 9-2.

Equivalent range	Number needle points to	Current value	Multiplier	Divisor
Example: 12 mA	8	8 mA	None	None
120 mA	2			
120 mA	7			
1.2 mA	4			
12 mA	9			
60 µA	5			
120 mA	7.5			
1.2 mA	3.2			
60 µA	6			
12 A	3			

Figure 9-4. Fill in the values.

7. Most multimeters are capable of measuring currents above 1 A. Up to 12 A of current can be measured with the meter illustrated. You will recall that the numerical value of the chosen range indicates the maximum value of current that can be measured on the range. For example, when the range selected is 12 mA, the meter is capable of measuring a maximum of 12 mA. Any current above 12 mA could damage the meter on this range. This is true for most current scales of most multimeters. If you are trying to measure an unknown current (no indication of its approximate value), you should start the measuring procedure by choosing the highest range on your meter and slowly adjusting the range downward until a current reading is indicated on the scale. Complete Figure 9-5 by indicating the proper meter range to be selected and the proper multiplier and divisor to be used with each value of current to be measured. (Note: Always select the smallest range that will allow you to measure the current.) Use your meter's ranges and scales for completing Table 9-5.

Current to be measured	Range to select	Multiplier	Divisor
Example: 0.9 mA	1 mA	0.1	10
2 A			
900 mA			
25 µA			
2 mA			
85 mA			
6 mA			
0.5 mA			
16 mA			
9 mA			
Unknown current			

Figure 9-5. Fill in the values.

8. Polarity of the meter is important when measuring current. The multimeter must always be connected in such a way as to allow the current to flow through the meter in the proper direction. Otherwise, the needle will deflect backwards, possibly causing internal damage to the meter. If the current is to flow through the meter in the proper direction, consideration must be given to the direction of the current. Current flows from negative to positive; thus, if the meter is connected correctly into a circuit, current will flow through the black test lead (–), through the meter, and then through the red test lead (+). The illustration of Figure 9-6 should be helpful.

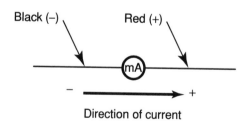

Direction of current

Figure 9-6.

9. In the illustrations of Figure 9-7, show which leads should be red (+) and which should be black (−) for the current meters indicated.

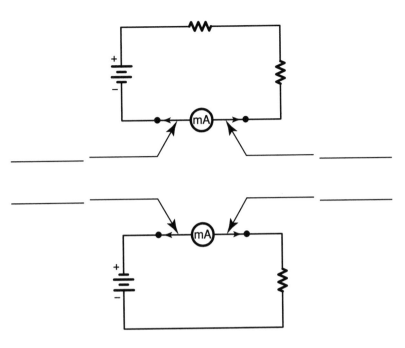

Figure 9-7.

10. Adjust the meter to the dc function, 120-mA range (or equivalent), and connect it into the circuit, as illustrated in Figure 9-8. Be sure to observe the proper meter polarity.

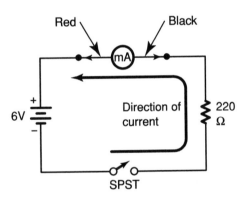

Figure 9-8.

11. Close the SPST switch and record the current indicated by the meter.

_____ mA, or _____ A.

12. Open the SPST switch and add the additional resistor illustrated in Figure 9-9.

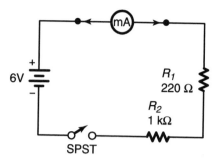

Figure 9-9.

13. Change the range of the multimeter to 12 mA (or equivalent), and close the switch. Record the current indicated.

_____ mA, or _____ A.

14. Disconnect the circuit of Figure 9-9 and connect the circuit of Figure 9-10.

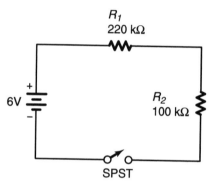

Figure 9-10.

15. Set the meter to the 60-µA range (or equivalent), and connect the meter to the circuit, observing proper polarity. Close the switch and record the current.

_____ µA, or _____ mA, or _____ A.

16. Disconnect the circuit illustrated in Figure 9-10.

17. Construct the circuit illustrated in Figure 9-11.

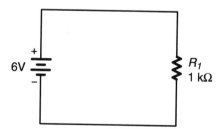

Figure 9-11.

18. Adjust the meter to the 12-mA range (or equivalent), and connect it into the circuit illustrated in Figure 9-11, observing proper polarity. Record the current.

_____ mA, or _____ A.

Activity 9—Measuring Current

Analysis

1. What is current? _____

_____

2. What do each of the dc ranges on the multimeter indicate? _____

_____

3. Describe how the multimeter is properly connected into a circuit to measure current. _____

_____

4. Why is proper polarity important to observe when one is measuring current? _____

_____

5. What meter ranges would be selected on your meter to measure the following currents?

Current	Range
0.009 A	_____
0.11 A	_____
0.8 A	_____
0.5mA	_____
0.000040 A	_____

6. What is the proper procedure to be followed when attempting to measure an unknown current? _____

_____

_____

7. What is the proper procedure to be followed when attempting to measure a current greater than 2 A? _____

_____

_____

Activity 10–Basic Electrical Problem Solving

Name _____ Date _____ Score _____

Objectives

Basic electrical problems are often encountered in any area that involves electrical control systems. The most basic problems in electrical systems involve Ohm's law. Ohm's law is a mathematical formula that explains the relationship between voltage, current, and resistance. This relationship must be understood before electrical oncepts are meaningful. In this activity, you will complete some practical problems by applying Ohms' law. Refer to Figure 10-1 for assistance.

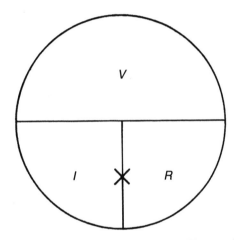

Figure 10-1. Ohm's law circle: V, voltage; I, current; R, resistance. To use the circle, cover the value you want to find and read the other values as they appear in the formula: $V = I \times R$, $I = V/R$, $R = V/I$.

Part 1–Ohm's Law Problems

1. A doorbell requires 0.2 amperes of current in order to ring. The voltage supplied to the bell is 120 volts. What is its resistance?

   $R =$ _____ ohms.

2. A relay used to control a motor is rated at 25 ohms resistance. What voltage is required to operate the relay if it draws a current of 0.25 ampere?

   $V =$ _____ volts.

3. An automobile battery supplies a current of 7.5 amperes to a headlamp with a resistance of 0.84 ohms. Find the voltage delivered by the battery.

   $V =$ _____ volts.

4. What voltage is needed to light a lamp if the current required is 2 amperes and the resistance of the lamp is 50 ohms?

   $V =$ _____ volts.

5. If the resistance of a radio receiver circuit is 240 ohms and it draws a current of 0.6 amperes, what voltage is needed?

$V =$ _____ volts.

6. A television circuit draws 0.15 amperes of current. The operating voltage is 120 volts. What is the resistance of the circuit?

$R =$ _____ ohms.

7. The resistance of the motor windings of an electric vacuum cleaner is 20 ohms. If the voltage is 120 volts, find the current drawn.

$I =$ _____ amperes.

8. The coil of a relay carries 0.05 amperes when operated from a 50-volt source. Find its resistance.

$R =$ _____ ohms.

9. How much current is drawn form a 12-volt battery when operating an automobile horn of 8 ohms resistance?

$I =$ _____ amperes.

10. Find the resistance of an automobile starting motor if it draws 90 amperes from the 12-volt battery.

$R =$ _____ ohms.

11. What current is drawn by a 5000-ohm electric clock when operated from a 120-volt line?

$I =$ _____ amperes.

12. Find the current drawn by a 50-ohm toaster from a 120-volt line.

$I =$ _____ amperes.

Part 2—DC Electrical Circuits

A fundamental understanding needed in electrical study is that of basic electrical circuits. All electrical circuits are classified as either series, parallel, or combination (series-parallel) circuits. In this activity, you will review each of these basic electrical circuits and complete a self-test on each type of circuit.

A. Series Circuit Problems—Self Test

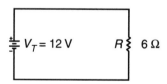

1. Find:

    Current through R (I_R) = _____ amperes.

    Voltage across R (V_R) = _____ volts.

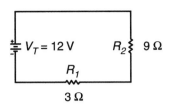

2. Find:

Total resistance (R_T) = _____ ohms.

Current through R_1 (I_{R_1}) = _____ amperes.

Current through R_2 (I_{R_2}) = _____ amperes.

Voltage across R_2 (V_{R_2}) = _____ volts.

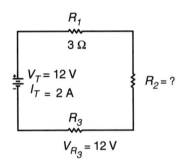

3. Find:

Voltage across R_1 (V_{R_1}) = _____ volts.

R_2 = _____ ohms.

Voltage across R_2 (V_{R_2}) = _____ volts.

R_3 = _____ ohms.

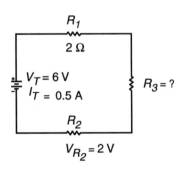

4. Find:

V_{R_1} = _____ volts.

R_2 = _____ ohms.

V_{R_3} = _____ volts.

R_3 = _____ ohms.

I_{R_2} = _____ amperes.

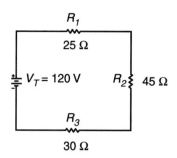

5. Find:

$I_T = $ _____ amperes.

$V_{R_1} = $ _____ volts.

$V_{R_2} = $ _____ volts.

$V_{R_3} = $ _____ volts.

Power converted by R_1

$(P_{R_1}) = $ _____ watts.

Power converted by R_2

$(P_{R_2}) = $ _____ watts.

Power converted by R_3

$(P_{R_1}) = $ _____ watts.

Total power converted by circuit

$(P_T) = $ _____ watts.

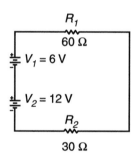

6. Find:

$V_T = $ _____ volts.

Total Power $(P_T) = $ _____ watts.

$R_T = $ _____ ohms.

$V_{R_1} = $ _____ volts.

$I_T = $ _____ amperes.

B. Parallel Circuit Problems—Self Test

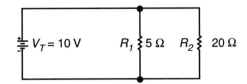

1. Find:

   Total resistance (R_T) = _____ ohms.

   Total current (I_T) = _____ amperes.

   Current through resistor R_1

   (I_{R_1}) = _____ amperes.

   Current through resistor R_2

   (I_{R_2}) = _____ amperes.

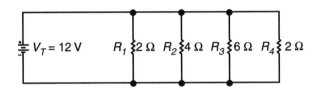

2. Find:

   Total resistance (R_T) = _____ ohms.

   Total current (I_T) = _____ amperes.

   Current through resistor R_1

   (I_{R_1}) = _____ amperes.

   Current through resistor R_2

   (I_{R_2}) = _____ amperes.

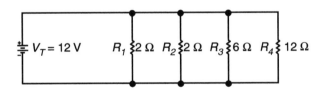

3. Find:

   Total resistance (R_T) = _____ ohms.

   Total current (I_T) = _____ amperes.

   Current through resistor R_1

   (I_{R_1}) = _____ amperes.

   Current through resistor R_2

   (I_{R_2}) = _____ amperes.

Current through resistor R_3

(I_{R_3}) = _____ amperes.

Current through resistor R_4

(I_{R_4}) = _____ amperes.

Total power (P_T) = _____ watts.

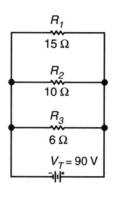

4. Find:

Total resistance (R_T) = _____ ohms.

Total current (I_T) = _____ amperes.

Current through resistor R_1

(I_{R_1}) = _____ amperes.

Current through resistor R_2

(I_{R_2}) = _____ amperes.

Current through resistor R_3

(I_{R_3}) = _____ amperes.

Power converted by resistor R_1

(P_{R_1}) = _____ watts.

Power converted by resistor R_2

(P_{R_2}) = _____ watts.

Power converted by resistor R_3

(P_{R_3}) = _____ watts.

5. Three resistors of 40 ohms, 90 ohms, and 75 ohms are connected in parallel across a 120-volt supply.

Find:

Total resistance (R_T) = _____ ohms.

Total current (I_T) = _____ amperes.

Current through each resistor:

I_{R_1} = _____ amperes.

I_{R_2} = _____ amperes.

I_{R_3} = _____ amperes.

C. Combination (Series-Parallel) Circuit Problems—Self Test

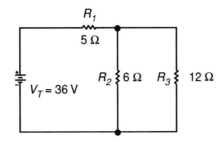

1. Find:

   Total resistance (R_T) = _____ ohms.

   Total current (I_T) = _____ amperes.

   Total power (P_T) = _____ watts.

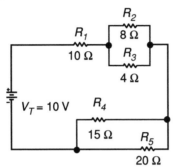

2. Find:

   Total resistance (R_T) = _____ ohms.

   Total current (I_T) = _____ amperes.

   Total power (P_T) = _____ watts.

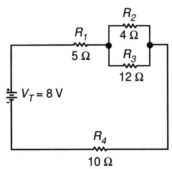

3. Find:

   Total resistance (R_T) = _____ ohms.

   Total current (I_T) = _____ amperes.

   Voltage across resistor R_1

   (V_{R_1}) = _____ volts.

   Voltage across resistor R_4

   (V_{R_4}) = _____ volts.

   Current through resistor R_2

   (I_{R_2}) = _____ amperes.

Current through resistor R_3

(I_{R_3}) = _____ amperes.

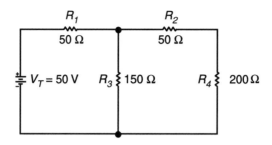

4. Find:

Total resistance (R_T) = _____ ohms.

Total current (I_T) = _____ amperes.

Voltage across resistor R_1

(V_{R_1}) = _____ volts.

Activity 11–Series DC Circuits

Name _____ Date _____ Score _____

Objectives

A *circuit* is defined as the complete path or paths through which current flows. All circuits must include a voltage source as well as conductors and/or components through which current flows. There are three broad classifications of all electrical circuits. These are series, parallel, and combination circuits.

The most easily understood circuit is the series circuit. The series circuit exhibits the following electrical characteristics:

- There is only one path for current.

- Current has the same value everywhere in the circuit.

- The voltage drops, when added, equal the source voltage.

- The total resistance of the circuit is determined by adding the values of all resistors in the circuit.
  In this activity, you will:

1. Examine the characteristics of series dc circuits.

2. Supply Ohm's law to series dc circuits.

3. Use a multimeter to make measurements in a series dc circuit.

Equipment and Materials

- Multimeter

- Variable dc power supply

- Resistors—100 Ω, 220 Ω, 200 Ω

- Connecting wires

Procedure

1. Construct the series circuit illustrated in Figure 11-1.

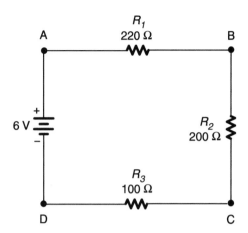

Figure 11-1. Series circuit.

2. In a series circuit, the total resistance is determined by adding the values of all the resistors. The formula normally used in determining the total resistance of a series circuit is $R_T = R_1 + R_2 + R_3 + \ldots + R_N$. This formula shows that when the values of resistors 1 (R_1), 2 (R_2), and 3 (R_3) are added together, the result is the total resistance. Compute the total resistance of the circuit illustrated in step 1.

   $R_T =$ _____ Ω (computed).

3. Prepare the meter to measure resistance.

4. Disconnect the power supply. Connect the meter in place of the power supply. Measure and record the total resistance of the circuit.

   $R_T =$ _____ Ω (measured).

5. How does the measured value of total resistance compare with the computed value? _____

   _____

6. Disconnect the meter and connect the power supply to its original location in the circuit.

7. Since you now know the total resistance of the circuit as well as the total voltage, use Ohm's law to compute the total current.

   $I_T =$ _____ mA.

8. Prepare the meter to measure direct current. (*Note:* The value of current computed in step 7 will help you select the proper range.)

9. Connect the meter at locations *A*, *B*, *C*, and *D* and record the currents at these points.

   Point A: _____ mA.   Point C: _____ mA.

   Point B: _____ mA.   Point D: _____ mA.

10. How do these currents compare? _____

    _____

11. Since the circuit in Figure 11-1 is a series circuit, the current is the same through each resistor. You should have seen this in step 9. Since you know the current through each resistor and the value of each resistor, use Ohm's law to compute the voltage drop across each resistor.

V across R_1 = _____ V.

V across R_2 = _____ V.

V across R_3 = _____ V.

12. Prepare the meter to measure dc volts. Measure and record the voltage drop across each resistor.

V across R_1 = _____ V.

V across R_2 = _____ V.

V across R_3 = _____ V.

13. How do the voltage drops computed in step 11 compare with the voltage drops measured in step 12? _____

14. Total the voltage drops in step 12.

Total voltage drop = _____ V.

15. How does this total voltage drop compare to the measured source voltage? _____

_____

Analysis

1. List the characteristics of a series circuit. _____

_____

_____

_____

2. What is the voltage drop across R_2 in the circuit of Figure 11-2? _____

_____

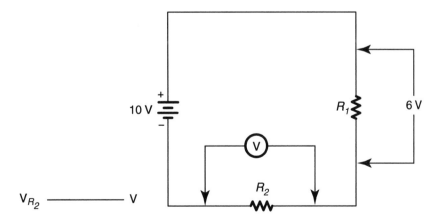

Figure 11-2. Computing voltage drop across a resistor.

3. If three resistors valued at 6 Ω, 8 Ω, and 10 Ω, respectively, were connected in series, what would be their total resistance?

R_T = _____ Ω.

4. In Figure 11-3, if 100 mA of current is flowing through R_1, how much current is flowing through R_2, R_3, and R_4? _____

_____

5. How many paths for current are illustrated in the circuit of Figure 11-3? _____

6. What is the total voltage drop in the circuit illustrated in Figure 11-3? _____

7. If three resistors valued at 10 Ω each were connected in series to a voltage source, what portion of the source voltage would appear across each resistor? _____

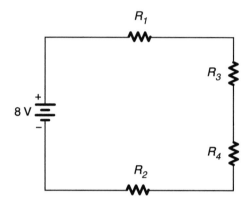

Figure 11-3. Four-resistor series circuit.

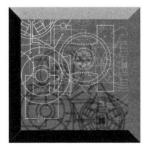

Activity 12–Parallel DC Circuits

Name _____ Date _____ Score _____

Objectives

Parallel circuits are circuits that provide two or more paths for current. Most of the light fixtures and wall outlets in your home are connected in parallel circuits. Many of the circuits that cause your radios and television sets to operate are parallel circuits.

The characteristics of parallel circuits are as follows:

- Two or more paths are provided for current.

- The voltage across each path or branch is the same.

- The sum of the currents in each path or branch will equal the total current.

- The total resistance (R_T) is computed using this formula:

$$R_T = \frac{1}{1/R_1 + 1/R_2 + 1/R_3 + \ldots 1/R_N}$$

where $R_1, R_2, \ldots R_N$ are the resistance of branches 1, 2, . . . N, respectively.

This equation is the same as
$1 / R_T = 1/R_1 + 1/R_2 + 1/R_3 + \ldots 1/R_N$

In this activity, you will:

1. Examine the characteristics of parallel dc circuits.

2. Apply Ohm's law to parallel dc circuits.

3. Use a multimeter to make measurements in a parallel dc circuit.

Equipment and Materials

- Multimeter

- Variable dc power supply

- Resistors—10 Ω, 15 Ω, 22 Ω

- Connecting wires

Procedure:

1. Construct the circuit illustrated in Figure 12-1. Use either a 1.5-V battery or a power supply.

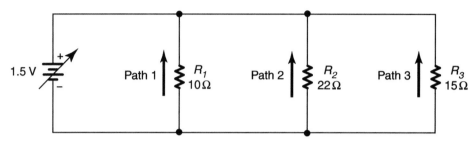

Figure 12-1. Three-branch parallel circuit.

2. How many different paths are provided for current? Is this a parallel or series circuit? _____

_____

3. What is the voltage across each path? _____

4. Compute the current in each path and the total current for the circuit.

Path 1 = _____ mA.

Path 2 = _____ mA.

Path 3 = _____ mA.

Total = _____ mA.

5. Compute the total resistance for the circuit illustrated in Figure 12-1.

R_T = _____ Ω.

6. Prepare the meter to measure direct current. Measure and record the currents through path 1 (R_1), path 2 (R_2), and path 3 (R_3), as well as the total current in the circuit. (The meter should be connected as illustrated in Figure 12-2 for each of these measurements. Thus the current in the path will correctly flow through the meter.)

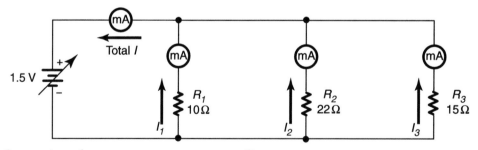

Figure 12-2. Connection of a current meter to measure direct current.

I_T = _____ mA.

I_1 = _____ mA.

I_2 = _____ mA.

I_3 = _____ mA.

7. How do the values of current in step 4 compare with the values of current in step 6? _____

_____

8. Disconnect the meter from the circuit and restore the circuit to its original condition as shown in Figure 12-1.

9. Prepare the meter to measure dc volts. Measure and record the voltage across each path of the circuit illustrated in Figure 12-1.

   Path 1 = _____ V.

   Path 2 = _____ V.

   Path 3 = _____ V.

10. How do these voltages compare to each other? _____

11. Disconnect the power supply. Prepare the meter to measure resistance. Connect the multimeter in place of the power supply and measure the total resistance of the circuit.

   R_T = _____ Ω.

12. How does the measured total resistance in step 11 compare with the computed total resistance in step 5? _____

_____

Analysis

1. List the characteristics of a parallel circuit. _____

_____

_____

_____

2. What is the current through R_3 in the circuit illustrated in Figure 12-3? _____

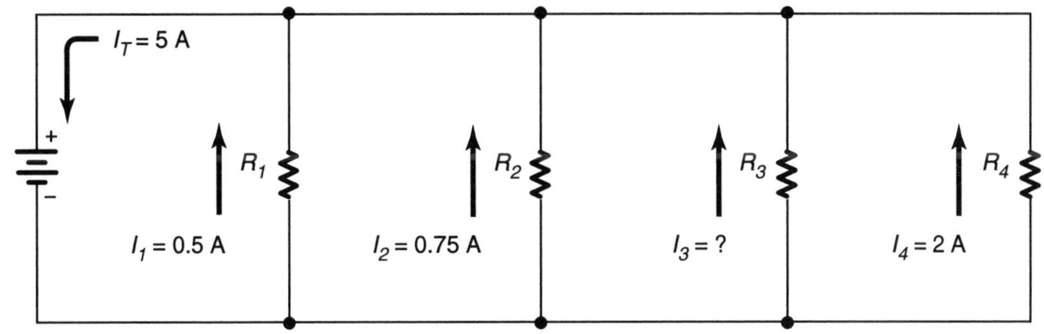

Figure 12-3. Circuit with an unknown branch current.

3. What is the total current in the circuit of Figure 12-4? _____

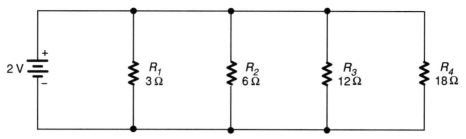

Figure 12-4. Circuit with an unknown total current.

4. What is the voltage across R_3 in the circuit of Figure 12-4? _____

5. What is the total resistance of the circuit shown in Figure 12-4? _____

6. If the resistance of each path of a parallel circuit is equal, how will the current in one path compare with the current in other paths? _____

7. In the circuit of Figure 12-5, indicate where the meter should be placed to measure total current. Use an X to indicate placement.

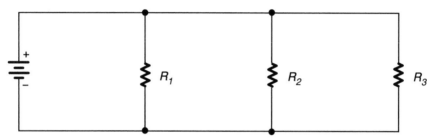

Figure 12-5. Current measurement.

8. In the circuit of Figure 12-5, indicate where the meter should be placed to measure current through R_2. Use a large dot to show the placement of the meter.

Activity 13–Combination DC Circuits

Name _____ Date _____ Score _____

Objectives

The combination circuit is the most used circuit in electronics. Its name is derived from the fact that it is actually a combination of the series and parallel circuits, since it exhibits the characteristics of both. When working with a combination circuit, you must treat its series portion as a series circuit and its parallel portion as a parallel circuit. You must also understand the influence of one portion of a combination circuit upon other portions.

All current, voltage, and power computations and measurements for this circuit are made in the same way as in previous circuits. The total resistance of a combination circuit is determined by first determining the series resistance, then the parallel resistance, and finally their sum.

In this activity, you will:

1. Examine the characteristics of a combination dc circuit.

2. Apply Ohm's law to combination dc circuits.

3. Use a multimeter to make measurements in a combination dc circuit.

Equipment and Materials

- Multimeter

- Variable dc power supply (or 6-V battery)

- Resistors—10 Ω, 15 Ω, 22 Ω

- Connecting wires

Procedure

1. Construct the circuit shown in Figure 13-1.

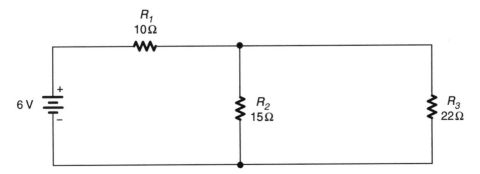

Figure 13-1. Series-parallel combination circuit.

2. Prepare the meter to measure dc voltage. Measure and record the voltages across R_1, R_2, and R_3.

$V_{R_1} = $ _____ V.

$V_{R_2} = $ _____ V.

$V_{R_3} = $ _____ V.

3. How did the voltage across R_2 compare with the voltage across R_3? _____

4. How does the sum of the voltage across R_1 and the voltage across R_2 or R_3 compare with the source voltage? _____

_____

5. Prepare the meter to measure direct current. Measure and record the currents through R_1, R_2, and R_3 and the total current.

$I_T = $ _____.

$I_{R_1} = $ _____.

$I_{R_2} = $ _____.

$I_{R_3} = $ _____.

6. How does the total current compare with the current through R_1? _____

_____

7. How does the sum of the currents through R_2 and R_3 compare with the total current? _____

_____

8. How does the sum compare with the current through R_1? _____

_____

9. Compute the total resistance for the circuit shown in Figure 13-1. (Remember to add the series resistance to the parallel resistance.)

$R_T = $ _____ Ω.

10. Disconnect the power supply from the circuit in Figure 13-1. Prepare the meter to measure resistance. Connect the meter where the power supply was connected. Measure and record the total resistance of the circuit.

$R_T = $ _____ Ω.

11. How did the measured resistance in step 10 compare with the computed resistance in step 9?

_____

12. Using Ohm's law, compute the total current for the circuit in step 1.

$I_T = $ _____ A.

13. How does the current computed in step 12 compare with the current measured in step 5? ____

_____

14. How does the total current in this circuit compare with the current through R_1? _____

_____

15. Compute the total parallel resistance of R_2 and R_3:

Parallel resistance of R_2 and $R_3 = $ _____ Ω.

16. In step 7, you found that the sum of the currents through R_2 and R_3 equaled the total current. Using the value of total current computed in step 12 and the parallel resistance computed in step 15, compute the voltage across the parallel resistors R_2 and R_3:

V across R_2 and R_3 = _____ V.

17. How does this voltage compute with the measured values in step 2? _____

_____

18. Using the total current computed in step 12, compute the voltage across R_1:

V_{R_1} = _____ V.

19. How does the computed voltage in step 18 compare with the measured voltage in step 2? ____

_____

20. How does the sum of the computed voltages across R_1, R_2, and R_3 compare with the source voltage? _____

_____

Analysis

1. How does the total current of a combination circuit compare with the current through its series components? _____

_____

2. How does the sum of the currents in the parallel paths of a combination circuit compare with the total current of the circuit? _____

_____

3. How does the sum of the currents in the parallel paths of a combination circuit compare with the current through the series components? _____

_____

4. How does the sum of the voltages across the series components and the parallel paths of a combination circuit compare with the source voltage? _____

5. If you subtracted the voltage across the parallel paths of a combination circuit from the source voltage, how would the results compare with the voltage across the series components of the circuit? _____

_____

6. If you subtracted the sum of the voltages across the series components of a combination circuit from the source voltage, how would the results compare with the voltage across the parallel paths of the circuit? _____

_____

7. If you added the sum of the resistance of the series components of a combination circuit to the parallel resistance of the circuit, how would the results compare with the total resistance of the circuit? _____

_____

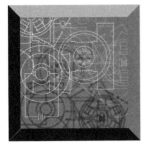

Activity 14–Power in DC Circuits

Name _____ Date _____ Score _____

Objectives

Electrical power is probably the electrical quantity most familiar to most people. All our electrical utility bills are computed relative to the amount of power consumed in a given amount of time. All electrical appliances have a specified power rating, which indicates how economically they can be operated. A knowledge of electrical power is important to anyone living in this age of energy conservation.

Electrical power is consumed each time a voltage causes current to flow. Power generally appears in the form of heat, light, or motion. The basic unit of measurement for electrical power is the watt. The milliwatt and kilowatt are also commonly used as units of measurement of power. Power is the product of current and voltage in a circuit. That means power (in watts) equals current (in amperes) multiplied by voltage (in volts). Actually, power can be computed when any two of the following electrical quantities are known: voltage (E), current (I), or resistance (R). The three basic formulas used in computing power in watts are:

$$P = IV \quad P = I^2R \quad P = V^2/R$$

where P, V, I, and R are in watts, volts, amperes, and ohms, respectively.

Many electrical components have an electrical power rating. Resistors are an excellent example, having common power ratings of 1/8 W, 1/4 W, 1/2 W, 1 W, and so on. This rating indicates the maximum current that can pass through the resistor without damaging the component.

In this activity, you will:

1. Calculate power values in dc circuits by applying the basic power formulas.

2. Use a meter to make measurements that are used to determine power in dc circuits.

Equipment and Materials

- Multimeter
- Variable dc power supply or 6-volt battery
- Lamp and socket, 6 volt
- Resistors—10 Ω, 100 Ω, 220 Ω, 1 kΩ
- Connecting wires
- SPST switch

Procedure

1. Construct the circuit shown in Figure 14-1.

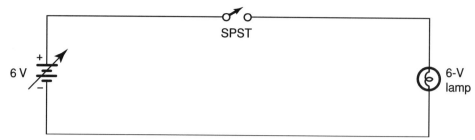

Figure 14-1. Simple six-volt lamp circuit.

2. Adjust the meter to measure current on the highest range.

3. Connect the meter to the circuit illustrated in Figure 14-1. Close the SPST switch and measure and record the current.

   Measured current = _____ mA, or _____ A.

4. Open the SPST switch, disconnect the meter, and restore the circuit to its original state, illustrated in Figure 14-1.

5. Prepare the meter to measure dc voltage.

6. Connect the meter to measure the voltage drop across the lamp illustrated in the circuit. Close the switch and record the voltage across the lamp.

   Measured voltage = _____ V.

7. You now have measured current through the lamp as well as the voltage across the lamp. Compute the power converted by the lamp.

   Computed power = _____ W.

8. An increase in power converted is caused when voltage increases and resistance remains the same. Notice the relationship between current, voltage, resistance, and power as you solve the following problems.

   a. $R = 10\ \Omega, I = 2$ A:

   $P =$ _____ W.

   $V =$ _____ V.

   b. $V = 100$ V, $I = 100$ mA:

   $P =$ _____ W.

   $R =$ _____ Ω.

   c. $I = 4$ A, $R = 5\ \Omega$:

   $P =$ _____ W.

   $V =$ _____ V.

   d. $R = 20\ \Omega, V = 10$ V:

   $P =$ _____ W.

   $I =$ _____ A.

e. $I = 1.5$ A, $V = 90$ V:

$P =$ _____ W.

$R =$ _____ Ω.

9. Construct the circuit shown in Figure 14-2.

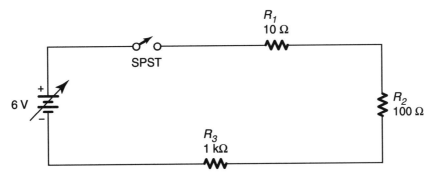

Figure 14-2. Circuit with three resistors.

10. Set the meter to measure direct current in the 12-mA range or equivalent.

11. Connect the meter to the circuit to measure current, close the SPST switch, and record the current indicated by the meter.

Measured current = _____ mA.

12. Compute the power converted by the circuit illustrated in Figure 14-2.

Computed power = _____W.

13. Open the switch, disconnect the meter, and alter the circuit to the one in Figure 14-3.

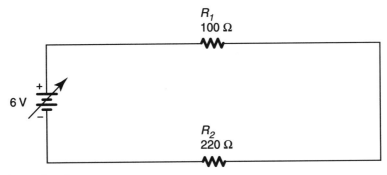

Figure 14-3. Two-resistor circuit.

14. Set the meter to the 12-V dc range. Measure and record the voltage across R_1, R_2, and R_3, respectively. Be sure to observe proper meter polarity.

V across R_1 = _____ V.

V across R_2 = _____ V.

V across R_3 = _____ V.

15. Compute the power converted by R_1 and R_2.

Computed power for R_1 = _____ V.

Computed power for R_2 = _____ V.

16. How does the power converted by R_1 compare with the power converted by R_2? Why do they differ? _____

_____

_____

Analysis

1. How much current flows through a 100-W light bulb when 100 V is connected to it? _____

2. How much power is converted by a 12-V lamp with a filament resistance of 2 Ω? _____

3. What is the maximum current that could safely pass through a 1000-Ω, 1-W resistor? _____

4. What power is converted by a 5-kΩ resistor that has 1 mA of current through it? _____

5. What power is converted by a light bulb that drops 6 V because of its 10-Ω filament? _____

6. What is the maximum voltage drop that could be developed by a 1.2-kΩ, 1/2-W resistor? _____

_____

_____

Activity 14—Power in DC Circuits

Activity 15–Basic AC Electrical Circuit Problems

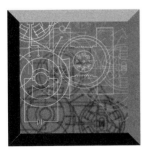

Name _____ Date _____ Score _____

Objectives

For more practice in solving ac circuit problems, you should complete this activity. It is important to have an understanding of basic ac circuit problems in order to solve more advanced circuit problems. These problems can be referred to when a similar problem is encountered in later activities. The organization of the activity is such that one example of each type of ac circuit calculation is done.

Part 1–Problems

1. Series inductance:
$L_T = L_1 + L_2 + L_3$.

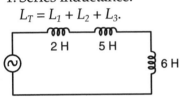

2 H   5 H   6 H

What is the inductance of this circuit?

$L =$ _____ henrys.

2. Parallel inductance:
$$\frac{1}{L_T} = \frac{1}{L_1} + \frac{1}{L_2} + \frac{1}{L_3}.$$

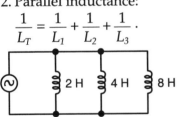

2 H   4 H   8 H

What is the inductance of this circuit?

$L =$ _____ henrys.

3. Inductive reactance: $X_L = 2\pi f L$.

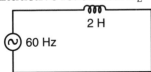

2 H

60 Hz

What is the inductive reactance?

$X_L =$ _____ ohms.

4. Series capacitance:
$$\frac{1}{C_T} = \frac{1}{C_1} + \frac{1}{C_2} + \frac{1}{C_3}.$$

F = farads
μF = microfarads (10^{-6})
pF = picofarads (10^{-12})

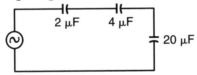

2 μF   4 μF   20 μF

What is the capacitance?

$C =$ _____ farads

5. Parallel capacitance:
$C_T = C_1 + C_2 + C_3$.

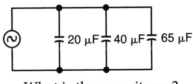

20 μF   40 μF   65 μF

What is the capacitance?

$C =$ _____ μF.

6. Capacitive reactance:
$$X_C = \frac{1}{2\pi f C}.$$

20 μF
60 Hz

What is the capacitive reactance?

$X_C =$ _____ ohms.

7. Impedance in series RC or RL circuit:

$Z = \sqrt{R^2 + (X_C)^2}$ or
$Z = \sqrt{R^2 + (X_L)^2}$

10 Ω $X_C = 10\ \Omega$

What is the impedance?

$Z =$ _____ ohms.

8. Impedance in series RLC circuits:

$$Z = \sqrt{R^2 + (X_L - X_C)^2}$$

30 Ω $X_C = 250\ \Omega$

$X_L = 210\ \Omega$

What is the impedance?

$Z =$ _____ ohms.

Part 2—Short Answer Items

Answer each of the following:

a. The opposition to current caused by inductance is called _____.

b. The letter symbol for inductive reactance is _____.

c. Inductive reactance is measured in _____.

d. The total opposition to current in an ac circuit is called _____.

e. The letter symbol for impedance is _____.

f. Impedance is measured in _____.

g. The relationship in electrical degrees of voltage and current in an ac circuit is called _____.

h. In a pure inductive circuit, current _____ voltage by _____.

i. In an ac circuit containing both resistance and inductance, current _____ voltage by _____ through _____.

j. Maximum effective voltage multiplied by maximum effective current equals_____ power.

k. Apparent power is measured in _____.

l. The power consumed by the resistance in an ac circuit is called the _____ power.

m. True power is measured in _____.

n. The ratio of true power to the apparent power is called _____.

o. Power factor is expressed as a _____ or a _____.

p. Apparent power multiplied by power factor equals _____.

Part 3—Basic AC Measurement Values

There are several basic ac values that are commonly used in electrical measurement and problem solving. These units are derived from ac sine wave relationships. The following ac sine wave values are often converted from one value to the other:

Effective value (rms) = 0.707 × peak value.

Average value = 0.636 × peak value.

Peak value = 1.41 × rms value.

Peak-to-peak value = 2.82 × rms value.

In this activity, you should review the preceding ac sine wave relationships and complete the self-test.

Activity 15—Basic AC Electrical Circuit Problems

Basic AC Values–Self Test

Solve the following problems by placing the correct answer in the blank space.

1. Compute the following effective values:

Peak voltage:

4 Vac = _____ V rms.

12 Vac = _____ V rms.

6 Vac = _____ V rms.

10 Vac = _____ V rms.

1.5 Vac = _____ V rms.

2 Vac = _____ V rms.

5 Vac = _____ V rms.

9 Vac = _____ V rms.

2. Compute the following peak and peak-to-peak (p-p) values:

Rms voltage:

3 Vac = _____ V peak; _____ V p-p.

8 Vac = _____ V peak; _____ V p-p.

7 Vac = _____ V peak; _____ V p-p.

9 Vac = _____ V peak; _____ V p-p.

10 Vac = _____ V peak; _____ V p-p.

15 Vac = _____ V peak; _____ V p-p.

11 Vac = _____ V peak; _____ V p-p.

18 Vac = _____ V peak; _____ V p-p.

Part 4–Trigonometry Problems for AC Circuits

In this activity, you will use a calculator to find sine, cosine, and tangent values. You will also find the values of angles when a sine, cosine, or tangent value is given. A knowledge of trigonometry is particularly important in dealing with ac circuits. Refer to Appendix B (Figure B-18) and Appendix C to complete this activity.

Abbreviations:

sin = sine
cos = cosine
tan = tangent
inv sin = angle whose sine is
inv cos = angle whose cosine is
inv tan = angle whose tangent is

Trigonometry Problems

Find the values of the following functions.

1. sin 138° = _____

2. sin 62° = _____

3. sin 212° = _____

4. cos 152° = _____

5. tan 141° = _____

6. sin 222° = _____

7. cos 230° = _____

8. tan 201° = _____

9. sin 318° = _____

10. cos 290° = _____

11. tan 310° = _____

12. inv sin 0.4226 = _____

13. inv cos 0.4848 = _____

14. inv tan 9.5144 = _____

15. inv tan 0.2309 = _____

16. inv sin 0.9613 = _____

17. inv cos 0.2924 = _____

18. inv sin -0.8660 = _____

19. inv cos -0.6820 = _____

20. inv tan -3.7321 = _____

Activity 16–Measuring AC Voltage

Name _____ Date _____ Score _____

Objectives

Alternating current is the most common form of current presently used in the United States. It is called alternating because it changes its direction periodically. The most frequently used unit of the time associated with ac is the second.

The number of ac cycles per second is known as frequency. Frequency is the number of times in 1 second that the ac signal moves from zero, reaches a peak in one direction, changes its direction, peaks in the opposite direction, and goes back to zero. These ac cycles per second (cps) are called hertz (Hz). The standard frequency of alternating current and voltage used in the United States is 60 Hz. The period of this standard frequency is 0.0166 seconds.

Since alternating current, voltage, and power are constantly changing, two types of electrical values are used in ac measurement. These are the instantaneous and the effective values. The instantaneous values are those used to describe the value of ac current, voltage, or power at any specified instant. The most common instantaneous value is the peak value: the maximum value of voltage, current, or power during any cycle. The effective values are more well known, since these values are used to describe the amount of voltage, current, and power that can be counted upon to produce light, heat, motion, or work of an electrical nature. The effective values of ac produce the same amount of work as dc values. Generally, the effective values are also called rms (root-mean-square) values.

Ohm's laws are used to compute ac values when the opposition to current is resistance only. Instantaneous values of voltage and current must be used to determine instantaneous power. Effective values must be used to determine effective power. The same procedure must be followed when determining voltage or current with a known resistance.

Instantaneous values of voltage, current, and power are converted to effective values, or vice versa, by using the mathematical constants 0.707 or 1.41 with the following formulas:

Effective value = 0.707 × peak value

Peak value = 1.41 × effective value

Peak-to-peak values (the distance from peak to peak) of alternating current and voltage are computed by using the following formula:

Peak-to-peak value = 2 × 1.41 × effective value

There is no peak-to-peak value for ac power.

In this activity, you will:

1. Study the characteristics of alternating current (ac).

2. Use a multimeter to measure ac voltage.

Equipment and Materials

- Multimeter
- DC power supply or 6-V battery
- AC source—30 Vac, center-tapped with potentiometer adjust
- SPST switch
- DPST switch
- Lamp with socket—6 volt
- Resistors—100 Ω, 220 Ω, 300 Ω
- Connecting wires

Procedure

1. Measure and record the resistance of the filament of the 6-V lamp.

   $R =$ _____ Ω.

2. Construct the circuit in Figure 16-1.

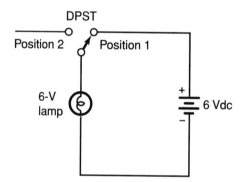

Figure 16-1.

3. Place the switch in position 1 and measure the dc voltage across the 6-V lamp. Record this voltage. _____ Vdc.

4. Place the switch in position 2 and disconnect the multimeter.

5. Connect a variable ac power supply to the circuit shown in Figure 16-2. (Note: Be sure to adjust the ac power supply to zero.)

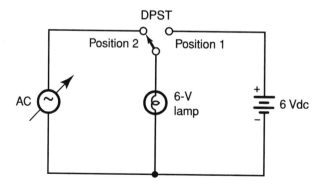

Figure 16-2.

If you do not have a variable ac power supply, variable ac voltage can be obtained as shown from the circuit of Figure 20-3.

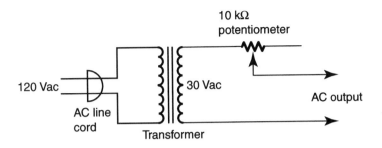

Figure 16-3.

6. Prepare the meter to measure ac voltage and connect it across the 6-V lamp.

7. Slowly adjust the variable ac voltage until the meter reads the same ac voltage as the dc voltage recorded in Step 3. Record this voltage. _____ Vac.

8. Slowly change the switch from position 2 to position 1 and back several times. How does this action affect the brightness of the bulb? _____

9. Using the data from Steps 1 and 3, compute the following dc values when the switch is in position 1.

Current through the lamp = _____ A.

Power dissipated by the lamp = _____ W.

10. Using the data from steps 1 and 7, compute the following ac values when the DPST switch is in position 2.

Current through the lamp = _____ A.

Power dissipated by the lamp = _____ W.

11. How did the dc data in step 9 compare with the ac data in Step 10? _____

_____

12. What conclusions can you reach when you compare ac effective values with identical dc values? _____

_____

_____

13. Disconnect the circuit shown in step 5 and construct the circuit in Figure 16-4.

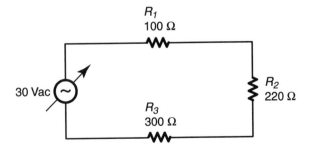

Figure 16-4.

14. Using a multimeter to measure ac voltage, complete the table in Figure 16-5. Use the proper formulas for computations.

Resistors	Measured RMS voltage (V) across resistor	Computed peak voltage across resistor = V × 1.41	Computed p-p voltage across resistor = V × 2.82	Computed RMS power (P) across resistor = $\frac{V^2}{R}$	Computed peak power across resistor = P × 1.41
R_1					
R_2					
R_3					

Figure 16-5. Complete the table.

Analysis

1. Define the following terms:

   a. Hertz: _____

   b. Frequency: _____

   c. Period: _____

   d. Cycle: _____

   e. Effective ac values: _____

   f. Instantaneous ac values: _____

2. Draw a sine wave in the space below. Indicate its four quadrants, positive peak, negative peak, and time base (in degrees).

3. Convert the following effective values to peak and peak-to-peak values.

   6 V rms = _____ V peak; _____ V p-p

   4 A rms = _____ A peak; _____ A p-p

   10 V rms = _____ V peak; _____ V p-p

   7 A rms = _____ V peak; _____ V p-p

Activity 16—Measuring AC Voltage

4. Convert the following peak values to effective values.

  12 W peak = _____ W rms

  100 mA peak = _____ mA rms

  2 mW = _____ mW rms

5. How does the power produced by 10 Vdc across 10 Ω compare with the power produced by 10 Vac rms across the same resistance? _____

6. Why is a cycle said to consist of 360°? _____

_____

7. Compute the periods for the following ac frequencies:

  10 kHz = _____ s

  1.2 kHz = _____ s

  1 kHz = _____ s

  0.6 kHz = _____ s

  60 kHz = _____ s

8. What is the standard frequency of the alternating current used in the United States? _____

9. What is alternating current? _____

_____

Activity 17–Inductance and Inductive Reactance

Name _____ Date _____ Score _____

Objectives

Inductance is an opposition to alternating current. It does not oppose direct current. Because ac encounters many forms of opposition not encountered by dc, all quantities that oppose ac are called *impedance*. The letter Z in mathematical formulas represents these opposing quantities. One of the impedance quantities that opposes ac is inductance.

Inductance is the characteristic of an ac circuit to oppose any increase or decrease in current. Inductance is represented by L in mathematical formulas and is measured in henrys (H). Physically, inductance is embodied in an inductor, which is no more than a coil. The impedance (Z) caused by an opposition other than resistance in ac circuits is known as *reactance*. In mathematical formulas, the letter X represents reactance. Reactance brought about by inductance is represented by X_L and is measured in ohms. The formula used to compute inductive reactance is:

$X_L = 2\pi fL$

where X_L is the inductive reactance in ohms, 2π is equal to 6.28, f is the frequency of ac in hertz, and L is the inductance in henrys.

In this activity, you will:

1. Examine the effects of ac on inductive reactance.

2. Make computations concerning total inductance in series and parallel.

Equipment and Materials

- Multimeter
- Audio signal generator
- Variable ac power source
- Inductor—4.5 H, iron-core (or any value from 2 to 15 H)
- Resistor—10 kΩ
- Switch (SPST)
- Lamp and socket—6 volt
- Connecting wires

Procedure

1. Measure and record the resistance of the inductor and 6-V lamp filament.

   Inductor dc resistance = _____ Ω.

   Lamp filament resistance = _____ Ω.

2. Construct the circuit in Figure 17-1.

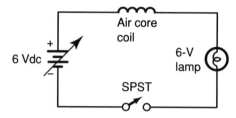

Figure 17-1.

3. Prepare a multimeter to measure dc voltage. Close the switch, measure and record the voltage across the lamp and the coil.

   V (coil) = _____ Vdc.

   V (lamp) = _____ Vdc.

4. How does the sum of the voltages across the coil and the lamp compare with the source voltage?

   Why? _____

   _____

5. Using the data from steps 1 and 3, compute the current through the coil and lamp.

   I = _____ A.

6. Replace the dc power supply with the ac power source as shown in Figure 17-2.

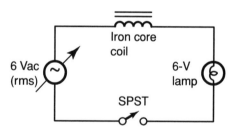

Figure 17-2.

7. Prepare the meter to measure ac voltage. Close the SPST switch. Measure and record the voltage across the lamp and coil.

   V (coil) = _____ Vac.

   V (lamp) = _____ Vac.

8. How does the sum of these voltages compare with the source voltage? Why? _____

   _____

9. How did the sum of the voltages in step 4 compare with the sum of the voltages in step 8? Why was this relationship different? _____

_____

10. Construct the circuit shown in Figure 17-3.

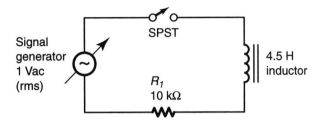

Figure 17-3.

11. Prepare the meter to measure ac voltage and connect it across R_1.

12. Close the switch and complete Figure 17-4. Adjust the signal generator to produce the ac frequencies indicated. (*Note:* The input voltage from the signal generator must be maintained at a constant level for each different frequency.)

AC frequency input	Voltage across R_1	Circuit current (I) $= \dfrac{V_{R_1}}{R_1}$	Inductive reactance (X_L) computed = $2\pi \bullet f \bullet L$
100 Hz			
200 Hz			
300 Hz			
400 Hz			
500 Hz			
1 kHz			
10 kHz			

Figure 17-4. Complete the table.

13. What is the relationship between ac frequency and circuit current? _____

_____

14. What is the relationship between ac frequency and inductive reactance? _____

_____

15. What is the relationship between inductive reactance and circuit current? _____

_____

Analysis

1. What factors determine the inductance of an inductor? _____

   _____

2. What factors determine the value of inductive reactance? _____

   _____

3. Why does inductance cause ac current to lag voltage? _____

   _____

4. What is the total inductance when inductors of 4 H and 3 H are connected in series with no
   mutual inductance factor? _____

   _____

5. What is the total inductance when inductors of 4 H and 3 H are connected in parallel with no
   mutual inductance factor? _____

   _____

6. Assume the inductors in Question 4 were connected to aid with a mutual inductance of 0.6 H.
   What is the total inductance? _____

   _____

7. Assume the inductors in Question 4 were connected to oppose with a mutual inductance of
   0.86 H. What is the total inductance? _____

   _____

8. Assume the inductors in Question 5 were connected to aid with a mutual inductance of 0.2 H.
   What is the total inductance? _____

   _____

9. Assume the inductors in Question 5 were connected to oppose with a mutual inductance of
   0.9 H. What is the total inductance? _____

   _____

10. What is mutual inductance? _____

   _____

11. If two inductors valued at 8 H and 10 H were connected in parallel, which would allow the
    most ac current if the ac frequency was 1000 Hz? _____

   _____

12. Why will an inductor oppose ac more than dc for any given voltage? _____

   _____

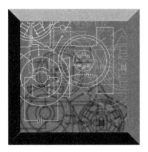

Activity 18–Capacitance and Capacitive Reactance

Name _____ Date _____ Score _____

Objectives

Capacitance is the property of an ac circuit to oppose any increase or decrease in voltage. It is present any time two conductors or plates are separated by a dielectric material. Capacitance is represented by C in mathematical formulas and is measured in farads (F). Components designed to add capacitance to a circuit are known as *capacitors* and are rated according to their capacitance in farads and their working voltage (maximum dc voltage that can be placed across a capacitor without doing damage to its dielectric). Capacitors, known for their ability to store an electrical charge, can be connected in series or parallel.

Capacitance in ac circuit causes current to *lead* the voltage. The impedance (Z) to alternating current due to capacitance is known as capacitive reactance (X_C) and is measured in ohms. Capacitive reactance is computed by using the following formula:

$$X_C = 1/2\pi fC$$

where X_C is the capacitive reactance in ohms, 2π is equal to 6.28, f is the frequency in hertz, and C is the capacitance in farads.

Capacitors can be further classified as electrolytic or nonelectrolytic. If electrolytic capacitors are used in a circuit, the polarity printed on the body of the capacitor *must be observed*.

In this activity, you will:

1. Study capacitance and capacitive reactance.

2. Examine how ac affects capacitive reactance.

3. Make simple computations concerning total capacitance in series and parallel.

Equipment and Materials

- Multimeter
- Signal generator
- Battery—6 volt
- Resistor—10 kΩ
- Capacitor—0.01 μF, 100 Vdc (working volts, dc)
- Switch (SPST)
- Connecting wires

Procedure

1. Prepare the multimeter to measure resistance in the ×1000 range or its equivalent. Measure and record the resistance of the 0.01-mF capacitor and the 10-kW resistor.

   Resistor resistance = _____ Ω.

   Capacitor resistance = _____ Ω.

2. How do these resistances compare? _____

   _____

3. From the data shown in step 1, which component would allow a direct current to flow and which would not? _____

   _____

4. Construct the circuit in Figure 18-1.

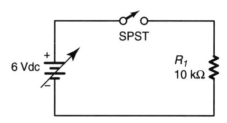

Figure 18-1.

5. Prepare the meter to measure direct current in the 10-mA range. Connect the meter into the circuit to measure current. Close the SPST switch and record the current.

   Current = _____ mA.

6. Alter the circuit in step 4 to the circuit shown in Figure 18-2.

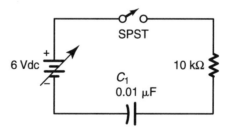

Figure 18-2.

7. Connect the meter to measure dc (in the 10-mA range). Close the switch and record the current.

   Current = _____ mA.

8. How did the current recorded in step 5 compare with the current recorded in step 7? _____

   _____

9. How do you account for the current difference in steps 5 and 7? _____

   _____

10. Alter the circuit in Figure 18-2 so it is like the one in Figure 18-3.

Activity 18—Capacitance and Capacitive Reactance

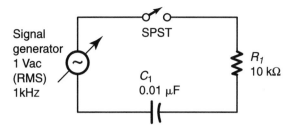

Figure 18-3.

11. Prepare the meter to measure ac volts in the 10-V range.

12. Connect the meter across R_1, close the SPST switch, and record the voltage.

   Voltage across R_1 = _____ Vac.

13. Using the resistance of R_1 as measured in step 1 and the voltage measured in step 12, compute the current in this circuit.

   I = _____ mA.

14. How did the current computed in step 13 compare with the current measured in step 7? _____

   _____

15. Since the two circuits are similar, how do you account for the difference in the currents in steps 7 and 13? _____

   _____

16. With the meter connected as in step 12, adjust the signal generator to the frequencies shown in Figure 18-4. Complete the figure for each frequency.

Frequency (Hz)	Measured voltage across R_1	Computed current $I = \dfrac{V_{R_1}}{R_1}$	Computed X_C
300			
500			
700			
900			
1100			
1300			
1500			
1700			
1900			
2500			
3000			

Figure 18-4. Complete the table.

17. From the data recorded in table of Figure 18-4, what is the relation between frequency, current, and capacitive reactance? As frequency increases, the voltage across R_1 _____, current _____, and X_L _____.

Analysis

1. What is capacitive reactance? _____

   _____

2. What is meant by a capacitor's working voltage? _____

   _____

3. What variables determine the capacitance of a capacitor? _____

   _____

4. If you connected in series two capacitors, valued at 4 μF each, what would be their total capacitance? _____

   _____

5. If you connected the two capacitors of Question 4 in parallel, what would be their total capacitance? _____

   _____

6. What is the relation between ac frequency and X_C? _____

7. How does capacitance affect the phase relation between alternating current and voltage? _____

   _____

8. How does the relation between ac frequency and X_C compare with the relation between ac frequency and X_L? _____

   _____

Activity 19–Series RL Circuits

Name _____ Date _____ Score _____

Objectives

Series RL circuits are encountered in many power systems. When an ac voltage is applied to this type of circuit, the current will be the same through each component. However, the voltage drops across each component will be distributed according to the relative value of resistance and inductive reactance in the circuit. The vector sum of the voltage drops will equal the source voltage. Therefore, in calculating voltage distribution in the circuit, we must use the following formula:

$$V_T = \sqrt{V_R^2 + V_L^2}$$

where, V_T is the total applied voltage, V_R is the voltage across the resistance, and V_L is the voltage across the inductance.

In this laboratory activity, you will observe the characteristics of a series ac circuit that has resistance and inductance. The inductance characteristic is associated with many power systems that have windings, such as motors, generators, and transformers.

Equipment and Materials

- Multimeter
- AC voltage source
- Resistor—10,000 Ω
- Inductor—8 H

Procedure

1. Construct the series RL circuit shown in Figure 19-1.

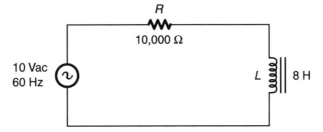

Figure 19-1. Series RL circuit.

2. Calculate the following:

  a. Inductive reactance $(X_L) = 2\pi fL =$ _____ Ω.

  b. Impedance $(Z) = \sqrt{R^2 + X_L^2} =$ _____ Ω.

3. With a meter, measure the following values:

  a. Voltage across the resistor $(V_R) =$ _____ volts ac.

  b. Voltage across the inductor $(V_L) =$ _____ volts ac.

  c. Current through the circuit $(I_T) =$ _____ amperes ac.

4. Experimentally determine the impedance of the circuit using the following method:

  a. $I_T = V_R/R =$ _____ A.

  b. $Z_T = V_T/I_T =$ _____ Ω.

5. Complete the impedance triangle of Figure 19-2 for the circuit you constructed by using the calculated values of R, X_L, and Z:

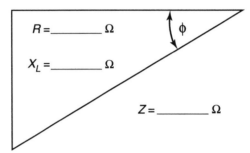

Figure 19-2. Impedance triangle.

6. Calculate the voltage drops in the circuit as:

  a. $V_R = I_T \times R =$ _____ volts ac.

  b. $V_L = I_T \times X_L =$ _____ volts ac.

7. Show that the voltages must be added vectorially by using the values you calculated:
$V_T = \sqrt{V_R^2 + V_L^2} =$ _____ volts ac.

8. Complete the following voltage triangles of Figure 19-3 for this circuit using the calculated, then the measured values of V_T, V_R, and V_L:

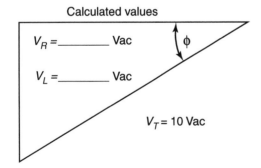
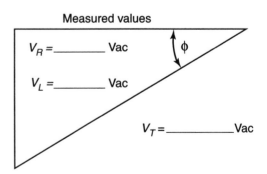

Figure 19-3. Voltage triangles.

9. Calculate the power converted in the circuit:

   a. True power (watts) = $I_T \times V_R$ = _____ watts.

   b. Apparent power (voltamperes) = $I_T \times V_T$ = _____ voltamperes.

   c. Reactive power (VAR) = $I_T \times V_L$ = _____ voltamperes-reactive.

10. Complete the power triangle of Figure 19-4 for the circuit using your calculated values of true power, apparent power, and reactive power.

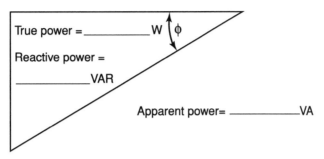

Figure 19-4. Power triangle.

Analysis

1. Why must the voltage in a series RL circuit be added vectorially? _____

_____

_____

2. Compare the calculated and measured values of the circuit using Figure 19-5.

Value	Calculated	Measured
I_T		
Z		
V_R		
V_L		

Figure 19-5. Comparison of calculated and measured values.

3. What are some factors that could cause a difference in your measured values and the calculated values? _____

_____

_____

4. Determine the phase angle of the circuit by using the following trigonometric methods:

   a. Phase angle (ϕ) = inv tangent X_L/R = _____°

   b. Phase angle (ϕ) = inv tangent V_L/V_R = _____°

   c. Phase angle (ϕ) = inv tangent VAR/W = _____°

5. Determine the power factor of the circuit as:

   Power factor (PF) = true power/apparent power = _____.

6. Determine the cosine of the phase angle (ϕ). This value should be equal to the current power factor (PF = cosine ϕ).

   Cosine (ϕ) = Cosine _____° = _____.

Activity 20–Series RC Circuits

Name _____ Date _____ Score _____

Objectives

Series RC circuits are used in many electrical systems. When an ac voltage is applied to this type of circuit, the characteristics are similar to those of a series RL circuit. You should recall that in a capacitive circuit, current leads voltage (leading phase angle) and in an inductive circuit, voltage leads current (lagging phase angle). Therefore, the vectorial relationships show the reactive values in the opposite direction.

In this laboratory activity, you will observe the characteristics of a series ac circuit that has resistance and capacitance.

Equipment and Materials

- AC voltage source
- Resistor—10,000 Ω
- Capacitor—1.0 µF
- Multimeter

Procedure

1. Construct the series RC circuit shown in Figure 20-1.

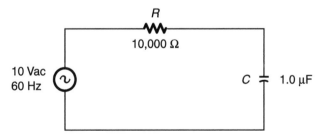

Figure 20-1. Series RC circuit.

2. Calculate the following:

   a. Capacitive reactance $(X_C) = 1/2\pi fC =$ _____ ohms.

   b. Impedance $(Z) =$ _____ $\sqrt{R^2 + X_C^2}$ ohms.

3. Measure the following values:

   a. Voltage across the resistor (V_R) = _____ volts ac.

   b. Voltage across the capacitor (V_C) = _____ volts ac.

   c. Current through the circuit (I_T) = _____ amperes ac.

4. Experimentally determine the impedance of the circuit using the following method:

   a. $I_T = V_R/R$ = _____ amperes.

   b. $Z_T = V_T/I_T$ = _____ ohms.

5. Complete the impedance triangle of Figure 20-2 for the circuit you constructed by using the calculated values of R, X_C, and Z.

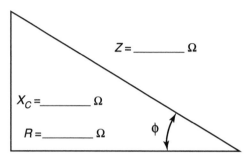

Figure 20-2. Impedance triangle.

6. Calculate the voltage drops in the circuit as:

   a. $V_R = I_T \times R$ = _____ volts ac.

   b. $V_C = I_T \times X_C$ = _____ volts ac.

7. Show that the voltages must be added vectorially by using the values you calculated:

   $V_T = \sqrt{V_R^2 + V_C^2}$ = _____ volts ac.

8. Complete the voltage triangles of Figure 20-3 for this circuit using the calculated, then the measured values of V_T, V_R, and V_C.

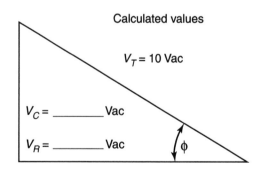

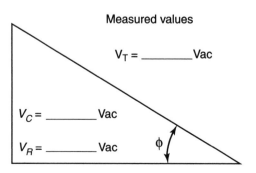

Figure 20-3. Voltage triangles.

Activity 20—Series RC Circuits

9. Calculate the power converted in the circuit.

  a. True power (watts) = $I_T \times V_R$ = _____ watts.

  b. Apparent power (voltamperes) = $I_T \times V_T$ = _____ voltamperes.

  c. Reactive power (VAR) = $I_T \times V_C$ = _____ voltamperes-reactive.

10. Complete the power triangle of Figure 20-4 for the circuit by using your calculated values of true power, apparent power, and reactive power.

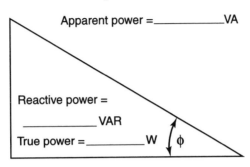

Apparent power = _____ VA

Reactive power = _____ VAR

True power = _____ W ϕ

Figure 20-4. Power triangle.

Analysis

1. Compare the calculated and measured values of the circuit in Figure 20-5.

Value	Calculated	Measured
I_T		
Z		
V_R		
V_C		

Figure 20-5. Comparison of calculated and measured values.

2. Determine the phase angle of the circuit by using the following trigonometric methods:

  a. Phase angle (ϕ) = inv tangent X_C/R = _____°

  b. Phase angle (ϕ) = inv tangent V_C/V_R = _____°

  c. Phase angle (ϕ) = inv tangent W/VA = _____°

3. Determine the power factor of the circuit as:

  Power factor (PF) = true power/apparent power = _____.

4. Determine the cosine of the phase angle. This value should equal the power factor of the circuit.

  Cosine ϕ = Cosine _____° = _____

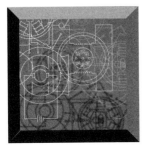

Activity 21–Series RCL Circuits

Name _____ Date _____ Score _____

Objectives

In this activity, you will observe the characteristics of an ac circuit that has resistive, capacitive, and inductive components. Since the effects of the capacitance and inductance are 180° out of phase with each other, we must analyze these circuits by using their vector relationships.

Equipment and Materials

- AC voltage source
- Resistor–10,000 Ω
- Capacitor–1.0 μF
- Inductor–8 H
- Multimeter

Procedure

1. Construct the series RCL circuit shown in Figure 21-1.

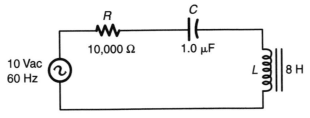

Figure 21-1. Series RCL circuit.

2. Calculate the following:
   a. Capacitive reactance $(X_C) = 1/2\pi fC =$ _____ Ω.
   b. Inductive reactance $(X_L) = 2\pi fL =$ _____ Ω.
   c. Total reactance $(X_T) = X_L - X_C =$ _____ Ω.
   d. Impedance $(Z) = \sqrt{R^2 + (X_L - X_C)^2} =$ _____ Ω.

3. With a meter measure the following values:

a. Voltage across the capacitor (V_C) = _____ volts ac.

b. Voltage across the inductor (V_L) = _____ volts ac.

c. Voltage across the resistor (V_R) = _____ volts ac.

d. Voltage across the circuit (V_T) = _____ amperes ac.

4. Experimentally determine the impedance of the circuit using the following method:

a. $I_T = V_R/R$ = _____ A.

b. $Z_T = V_T/I_T$ = _____ Ω.

5. Complete the impedance triangle of Figure 21-2 for the circuit you constructed by using the calculated values of R, X_L, and Z.

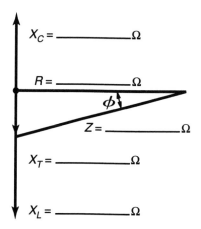

Figure 21-2. Impedance triangle.

6. Calculate the voltage drops in the circuit as:

a. $V_R = I_T \times R$ = _____ volts ac.

b. $V_C = I_T \times X_C$ = _____ volts ac.

c. $V_L = I_T \times X_L$ = _____ volts ac.

d. $V_X = V_L \times V_C$ = _____ volts ac.

7. Show that the voltages must be added vectorially by using the values you calculated.

$V_T = \sqrt{V_R^2 + (V_L - V_C)^2}$ = _____ volts ac.

8. Complete the voltage triangles of Figure 21-3 circuit using the calculated, then the measured values of V_T, V_R, V_L, and V_C.

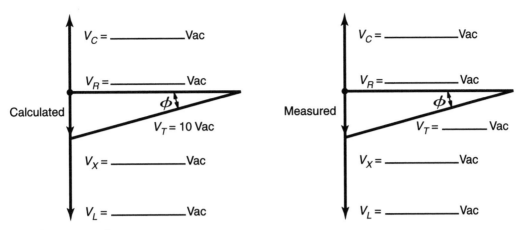

Figure 21-3. Voltage triangles.

9. Calculate the power converted in the circuit:
   a. True power = $I_T \times V_R$ = _____ W.
   b. Apparent power = $I_T \times V_T$ = _____ VA.
   c. Capacitive reactive power = $I_T \times V_C$ = _____ VAR.
   d. Inductive reactive power = $I_T \times V_L$ = _____ VAR.
   e. Total reactive power = $I_T \times V_X$ = _____ VAR.

10. Complete the power triangle of Figure 21-4 for the circuit by using your calculated values of true power, apparent power, and reactive power:

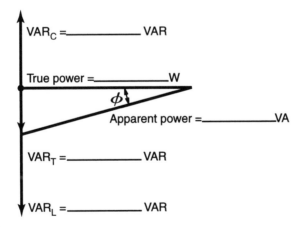

Figure 21-4. Power triangle.

Analysis

1. Compare the calculated and measured values of the circuit in Figure 21-5.

Value	Calculated	Measured
I_T		
Z		
V_R		
V_C		
V_L		

Figure 21-5. Comparison of calculated and measured values.

2. Determine the phase angle of the circuit by using the following trigonometric methods:

   a. Phase angle (ϕ) = inv tangent X_T/R = _____ °.

   b. Phase angle (ϕ) = inv tangent V_X/V_R = _____ °.

   c. Phase angle (ϕ) = inv tangent VAR_T/W = _____ °.

3. Determine the power factor of the circuit as:

   Power factor (PF) = true power/apparent power = _____ = _____%.

4. Determine the cosine of the phase angle (ϕ). This value should equal the power factor of the circuit.

   Cosine (ϕ) = Cosine _____ ° = _____.

5. How does the circuit of this experiment show power factor correction? _____

   _____

6. As total reactive power increases, the phase angle (increases/decreases). _____

Activity 22–Parallel RL Circuits

Name _____ Date _____ Score _____

Objectives

Another category of basic ac circuits is the parallel type. Parallel circuits are employed for many industrial control applications. The basic calculations necessary to study parallel ac circuits are somewhat different from those of series circuits. Since the impedance (Z) of a parallel circuit is less than the individual values of resistance, inductive reactance, or capacitive reactance, the impedance relationships must be modified. Also, a vector relationship exists between the currents that flow in the branches of the circuit.

In this activity, you will observe the impedance and current relationships of a parallel resistive-inductive (RL) circuit with ac applied to it.

Equipment and Materials

- AC voltage source
- Resistor—1000 Ω
- Inductor—8 H
- Multimeter

Procedure

1. Construct the parallel RL circuit shown in Figure 22-1.

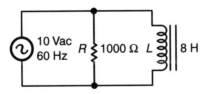

Figure 22-1. Parallel RL circuit.

2. Calculate the following:

   a. Inductive reactance $(X_L) = 2\pi f L =$ _____ Ω.

   b. Current through the resistor $(I_R) = V/R =$ _____ A.

   c. Current through the inductor $(I_L) = V/X_L =$ _____ A.

   d. Total current $(I_T) = \sqrt{I_R^2 + I_L^2}$ = _____ A.

   e. Impedance $(Z) = V/I_T =$ _____ Ω.

3. Complete the current triangle of Figure 22-2 for this circuit using the values from Step 2.

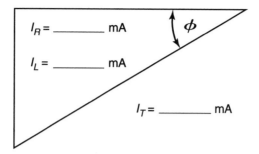

Figure 22-2. Current triangle.

4. With a meter, measure the following values:
   a. Current through the resistor (I_R) = _____ A.
   b. Current through the inductor (I_L) = _____ A.
   c. Total current (I_T) = _____ A.

5. Since the impedance is less than the resistance or inductive reactance of the circuit, an *admittance* diagram must be used. In this diagram, admittance $(Y) = 1/Z$, conductance $(G) = 1/R$, and inductive susceptance $(B_L) = 1/X_L$. The unit of measurement is the mho or siemens (S). Note that "mho" is "ohm" spelled backward.

6. Calculate the values necessary to complete the admittance triangle of Figure 22-3.

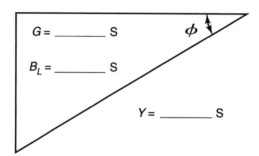

Figure 22-3. Admittance triangle.

   a. Admittance $(Y) = 1/Z$ = _____ S.
   b. Conductance $(G) = 1/R$ = _____ S.
   c. Inductive susceptance $(B_L) = 1/X_L$ = _____ S.

7. Calculate the power converted in the circuit:
   a. True power (watts) = $I_R^2 \times R$ = _____ W.
   b. Apparent power (voltamperes) = $I_T^2 \times Z$ = _____ VA.
   c. Reactive power (VAR) = $I_L^2 \times X_L$ = _____ VAR.

8. Complete the power triangle of Figure 22-4 for the circuit:

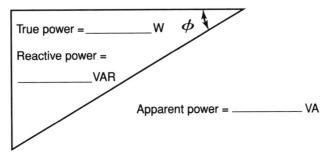

True power = _____ W

Reactive power =

_____ VAR

Apparent power = _____ VA

Figure 22-4. Power triangle.

Analysis

1. Compare the calculated and measured values of the circuit in Figure 22-5.

Value	Calculated	Measured
I_R		
I_L		
I_T		

Figure 22-5. Comparison of calculated and measured values.

_____

_____

2. Determine the phase angle of the circuit by the following methods:

   a. Phase angle (ϕ) = inv cosine I_R/I_T = _____°.

   b. Phase angle (ϕ) = inv sine VAR/VA = _____°.

3. Determine the power factor of this circuit.

   Power factor (PF) = true power/apparent power = _____.

4. Compare the power factor of this circuit with the value of cosine (ϕ). _____

_____

5. Define the following terms associated with parallel ac circuits:

   a. Admittance:

      _____

      _____

   b. Conductance:

      _____

      _____

   c. Inductive susceptance:

      _____

      _____

   d. Capacitive susceptance:

      _____

      _____

6. What are some sources of error that account for some difference between calculated and measured values in this activity? _____

   _____

   _____

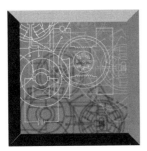

Activity 23–Parallel RC Circuits

Name _____ Date _____ Score _____

Objectives

In this laboratory activity, you will observe the characteristics of a parallel RC circuit with ac applied. This type of circuit is similar to the RL circuit. However, the effect of capacitance is opposite that of inductance. Similar vector relationships exist in this circuit.

Equipment:

- Multimeter
- AC voltage source
- Resistor—1000 Ω
- Capacitor—1.0 µF

Procedure

1. Construct the parallel RC circuit shown in Figure 23-1.

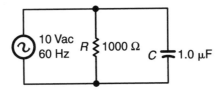

2. Calculate the following:
   a. Capacitive reactance $(X_C) = 1/2\pi fC =$ _____ Ω.
   b. Current through the resistor $(I_R) = V/R =$ _____ A.
   c. Current through the capacitor $(I_C) = V/X_C =$ _____ A.
   d. Total current $(I_T) = \sqrt{I_R^2 + I_C^2} =$ _____ A.
   e. Impedance $(Z) = V/I_T$ _____ $= \Omega$.

3. Complete the following current triangle of Figure 23-2 for this circuit using the values from Step 2.

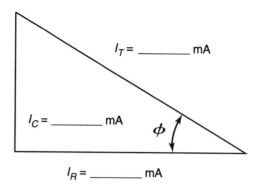

I_T = _____ mA

I_C = _____ mA

ϕ

I_R = _____ mA

4. Measure the following ac current values:
   a. Current through the resistor (I_R) = _____ A.
   b. Current through the capacitor (I_C) = _____ A.
   c. Total current (I_T) = _____ A.

5. Calculate the following values:
   a. Admittance (Y) = $1/Z$ = _____ S.
   b. Conductance (G) = $1/R$ = _____ S.
   c. Capacitive susceptance (B_C) = $1/X_C$ = _____ S.

6. Complete the admittance triangle of Figure 23-3 for the circuit:

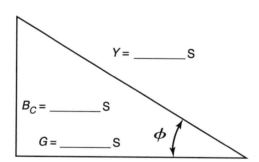

Y = _____ S

B_C = _____ S

G = _____ S

ϕ

7. Calculate the power converted in the circuit:
   a. True power = $I_R^2 \times R$ = _____ W.
   b. Apparent power = $I_T^2 \times Z$ = _____ VA.
   c. Reactive power = $I_C^2 \times X_C$ = _____ VAR.

8. Complete the power triangle of Figure 23-4 for the circuit:

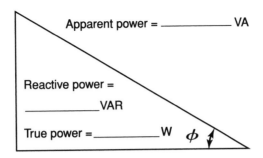

Apparent power = _____ VA

Reactive power = _____ VAR

True power = _____ W ϕ

9. This concludes the experiment.

Analysis

1. Insert the calculated and measured values for the circuit of this experiment into the following table. Compare the values.

Value	Calculated	Measured
I_R		
I_C		
I_T		

2. Determine the phase angle (ϕ) of the circuit using the following methods:

   a. Phase angle (ϕ) = inv sine I_C/I_T = _____°.

   b. Phase angle (ϕ) = inv tangent VAR/W = _____°.

3. Determine the value of:

   a. Cosine (ϕ) = cosine _____° = _____.

   b. Power factor = true power/apparent power = _____ = _____%.

4. If an 8-henry inductor was connected in parallel with the circuit of this activity, compute the following:

   a. $X_L = 2\pi fL$ = _____ Ω.

   b. V/X_L = _____ A.

   c. $I_X = I_L - I_C$ = _____ A.

   d. $I_T = \sqrt{I_R^2 + I_X^2}$ = _____ A.

   e. V/I_T = _____ Ω.

   f. ϕ = inv sine (I_X/I_T) = _____°.

   g. PF = cosine ϕ = cosine _____° = _____ _____%.

5. How do the values of power factor for the original circuit and that of the preceding differ? Why? _____

_____

_____

6. How do the values of total current differ in the two circuits? Why? _____

_____

_____

Activity 24—Working with Permanent Magnets

Name _____ Date _____ Score _____

Objectives

In this activity, you will learn the properties of magnetic fields. It is important to understand the effects of magnetism since the operation of many types of machines depends on magnetic fields.

Equipment and Materials

- Permanent magnets (2)
- Iron filings
- Paper

Procedure

1. Obtain two permanent magnets, a sheet of paper, and some iron filings.
2. Place one magnet on a flat surface with a sheet of paper placed over it.
3. Carefully pour iron filings onto the paper above the magnet.
4. Carefully lift the paper and place the second magnet under the paper with the two north poles about one inch apart. Make a sketch of the magnetic field pattern on a sheet of paper.
5. Place the north and south poles about one inch apart. Make a sketch of the magnetic field patterns.
6. Return all equipment.

Analysis

1. Discuss the laws of magnetism with the class.

Activity 25—The Nature of Magnetism

Name _____ Date _____ Score _____

Objectives

Magnetism is one of the longest-known natural forces. It was first discovered and used by ancient cultures as a curiosity. Many believed that this force was magic and therefore to be feared. The first magnets used were natural magnets called lodestones. They were first put to practical use in navigation. Someone discovered that when these devices were suspended by a string and allowed to move freely, they would always align themselves to a point to the north. Thus, natural magnetism was first used for compasses.

Much later it was discovered that magnetism could be used to create an electric current, and an electric current could be used to create a magnetic field. This relationship makes a knowledge of magnetism extremely important.

In this activity, you will examine the characteristics of natural magnetism.

Equipment and Materials

- Permanent magnets (2)
- Magnetic compass

Procedure

1. Place a permanent magnet on the surface in front of you.

2. Examine the needle of the compass. You will find that the north-seeking pole (south pole) of the compass needle is painted or otherwise colored to distinguish it from the north pole.

3. Use the compass to identify the north pole and south pole of the permanent magnet. (Remember, like poles repel; unlike poles attract.) Mark the north and south poles of your permanent magnet.

4. If the north pole of one magnet is brought near the north pole of another magnet, what happens? Why? _____

   _____

5. If the south pole of one magnet is brought near the south pole of another magnet, what happens? Why? _____

   _____

6. If the north pole of one magnet is brought near the south pole of another magnet, what happens? Why? _____

   _____

Analysis

1. Describe the reaction of like poles and unlike poles of magnetic fields. _____

_____

2. Why can a compass be used to detect the presence of a magnetic field? _____

_____

3. Define the following terms:

   a. Flux: _____

   b. Magnetic field: _____

   c. Magnetic lines of force: _____

4. In what direction does magnetic flux travel externally around a magnet? _____

5. Sketch the magnetic field of a permanent magnet in the space that follows.

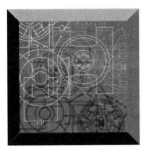

Activity 26–Making an Electromagnet

Name _____ Date _____ Score _____

Objectives

In this activity, you will learn the properties of an electromagnet by making one that will actually work. It is important to understand the effects of electromagnetism since the operation of many types of machines depends on electromagnetic fields.

Equipment and Materials

- Bolt—3/8- to 3/4- in. in diameter and about 4 in. long

- Solid copper wire—10 to 20 feet of No. 18 to No. 24 diameter

- Lantern battery—6 volt

- Compass

Procedure

1. Use the bolt as an iron core. Start at one end and wind the wire around the core in the *same direction*. Leave 6 to 10 in. of wire on each end of the wire, for connecting the battery. Wrap tape around the wire to secure it in place. The finished coil should have between 100 and 200 turns of wire.

2. Momentarily connect the ends of the coil to a 6-V lantern battery. It may be necessary to remove about one-half inch of insulation from the ends of the wire. This will allow good electrical contact to be made. Do *not* keep the battery connected for over a few seconds at a time.

3. Test the electromagnet with a compass to see if it has a north polarity at one end and a south polarity at the other end.

4. Try picking up metal objects with the electromagnet.

5. Let your instructor check your electromagnet.

6. Return all materials.

Analysis

Discuss the construction and operation of an electromagnet. Be sure to discuss ways to make the electromagnet stronger.

_____

_____

_____

_____

_____

_____

Activity 27–Problems for Electrical Distribution Systems

Name _____ Date _____ Score _____

Problems

1. A 1500-turn primary winding has 120 volts applied. The secondary voltage is 20 volts. How many turns of wire does the secondary winding have?

2. When a 200-ohm resistance is placed across a 240-volt secondary winding, the primary current is 30 amperes. What is the value of the primary voltage of the transformer?

3. A transformer has a 20:1 turns ratio. The primary voltage is 4800 volts. What is the secondary voltage?

4. A transformer has a 2400-volt primary winding and a 120-volt secondary. The primary winding has 1000 turns. How many turns of wire does the secondary winding have?

5. A 240-volt to 4800-volt transformer has a primary current of 95 amperes and a secondary current of 4 amperes. What is its efficiency?

6. A transformer has a 20-kVA power rating. Its primary voltage is 120 volts and its secondary voltage is 480 volts. What are the values of its maximum primary current and secondary current?

7. Convert the following diameters of electrical conductors into mils.

   a. 1/8 inch.

   b. 1/4 inch.

   c. 3/8 inch.

   d. 1/2 inch.

   e. 0.2 inch.

   f. 0.28 inch.

   g. 0.15 inch.

   h. 0.66 inch.

8. Calculate the cross-sectional area of the following electrical conductors (in circular mils).

   a. 3/16 inch diameter.

   b. 0.5 inch square conductor.

   c. 0.38 inch diameter.

   d. 1/2 inch x 3/8 inch rectangular conductor.

9. Calculate the resistance of 480 feet of copper conductor that is 0.28 inches in diameter.

10. Use the proper *formula* to calculate the resistances of the following electrical conductors.

    a. 380 feet of No. 10 AWG copper.

    b. 2500 feet of No. 000 AWG copper.

    c. 230 feet of 500 MCM copper.

    d. 1900 feet of No. 8 AWG aluminum.

    e. 220 feet of No. 10 AWG aluminum.

11. Use the proper *table* to calculate the resistance of the following electrical conductors.

    a. 600 feet of No. 0000 AWG copper.

    b. 500 feet of No. 1 AWG aluminum.

    c. 300 feet of No. 12 AWG copper.

    d. 200 feet of 250 MCM aluminum.

    e. 1800 feet of No. 10 AWG copper.

12. In order to carry the following currents, what are the minimum sizes of conductors needed? No more than three conductors will be used in the same raceway.

    a. 380 amperes using THW copper.

    b. 95 amperes using THW aluminum.

    c. 180 amperes using RH copper.

    d. 230 amperes using TW copper.

    e. 230 amperes using R copper.

    f. 650 amperes using RHW copper.

    g. 580 amperes using T copper.

    h. 82 amperes using TH aluminum.

13. Find the equivalent size of THW aluminum conductors for the following copper conductors.

    a. 500 MCM, TW.

    b. 750 MCM, TH.

    c. No. 00 AWG, R.

    d. No. 4 AWG, T.

    e. No. 8 AWG, RHW.

14. Three 20,000-watt loads are connected to a three-phase, wye-connected system. The power factor of the system is 0.9. Find the minimum size of RH copper feeder conductors needed to supply this load.

15. Determine the branch-circuit rating and size of conductors needed for the following systems:

    a. 2500-watt, 240-volt heater.

    b. 1800-watt, 120-volt washer.

    c. 12,000-watt, 240-volt range.

    d. 800-watt, 120-volt toaster.

16. Calculate the maximum distance a 120-volt, 30-ampere branch circuit can extend from the power source when a 3-kW appliance is connected to the circuit. The voltage drop should be limited to 2%.

17. A 120-volt 20-ampere branch circuit (using No. 12 copper conductors) extends 150 feet. Loads connected at 30-foot intervals each draw 3.5 amperes. With all loads connected, calculate the voltage at the last outlet.

18. Calculate the longest distance a single-phase 20-ampere, 120-volt branch circuit can extend from the power source. Use No. 12 AWG copper conductors and limit the voltage drop to 2%.

19. A single-phase 120-volt load is rated at 40 kilowatts. The feeder circuit will be 400 feet of RH copper conductors. Find the minimum conductor size necessary to supply this load and still limit the voltage drop to 1%.

20. A 240-volt three-phase delta system converts 30 kW of power per phase (balanced). The system power factor is unity (1.0). The feeders will be 400 feet of RHW copper conductor. Find the minimum conductor size necessary to limit the voltage drop to 2%.

Activity 28–Transformer (Self Test)

Name _____ Date _____ Score _____

Procedure

Complete the following by providing the proper responses.

Identify the parts of the transformer shown in Figure 28-1.

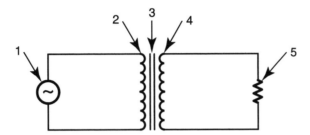

Figure 28-1.

1. _____

2. _____

3. _____

4. _____

5. _____

6. What is the secondary voltage of the transformer shown in Figure 28-2?

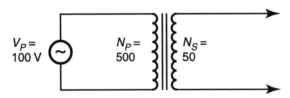

Figure 28-2.

$V_S =$ _____.

7. If the primary current of the transformer shown in Figure 28-2 is 5 A, what would the secondary current equal?

$I_S =$ _____.

8. A transformer that only has one winding is called _____.

9. The major reason for using transformers to increase voltage for power transmission is to reduce _____.

10. An electrical device consisting of two coils very close together yet electrically insulated from each other is a(n) _____.

11. A transformer coil connected to a source of ac voltage is the _____ winding.

12. A transformer coil connected to the load is the _____ winding.

13. If there are less turns in the secondary winding than in the primary winding, the device is a(n) _____ transformer.

14. If there are more turns in the secondary winding than in the primary winding, the device is a(n) _____ transformer.

15. To reduce the opposition to magnetic lines of force between the primary and secondary, the windings may be wound on a(n) _____ _____.

16. If 120 V are applied to a primary of a step-down transformer with a 10:1 turns ratio, the voltage induced in the secondary is _____ V.

17. If 120 V are applied to a primary of a step-up transformer with a 1:2 turns ratio, the voltage induced in the secondary is _____ V.

18. If the voltage across the complete secondary winding of a center-tap transformer is 240 V, the voltage from one outside conductor to the center tap is _____ V.

19. A transformer with more than one secondary winding is called a(n) _____ transformer.

20. The ratio of power output to power input of a transformer is called _____.

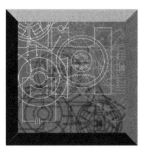

Activity 29–Transformer Operation

Name _____ Date _____ Score _____

Objectives

In this activity, you will learn the properties of a transformer by checking the operation of a working transformer. It is important to understand the operation of transformers since they are used in power distribution systems.

Equipment and Materials

- Multimeter
- Transformer

Procedure

1. Obtain a transformer and a multimeter. Record all available data on your transformer on a sheet of paper.

2. Identify the primary windings of the transformer. These primary windings will be connected to an ac voltage source. Do *not* plug it in until you are ready to make the necessary measurements. Remove the power cord when the transformer is not in use. Be very aware of safety. It is best to make measurements with a low ac voltage applied to the primary (from 6 to 15 V). *If a 120-Vac source is used, be very careful.*

3. Some transformers have one secondary winding. Many transformers have three or more secondary windings. Count the number of secondary windings on the transformers.

   Number of secondary windings = _____.

4. Make sure that none of the wires from the secondary windings touch.

5. Prepare the meter to measure ac voltage. Begin with the highest ac voltage range for each secondary voltage measurement.

6. Carefully connect the primary winding to an ac source. Measure the voltage across each secondary winding. Record these voltages.

7. Disconnect the ac voltage from the transformer primary. With no voltage applied, make the following resistance measurements with a multimeter.

   a. Resistance of primary winding = _____ Ω.

   b. Resistance of each secondary winding = _____ Ω.

Analysis

Discuss your readings and the basic operation of a transformer.

_____

_____

_____

_____

_____

Activity 30–Transformer Turns Ratio

Name _____ Date _____ Score _____

Procedure

1. Review the diagrams shown in Figure 30-1. Observe there are 100-turn windings and 500-turn windings shown in various combinations with 10 volts ac applied to the primary winding of each illustration.

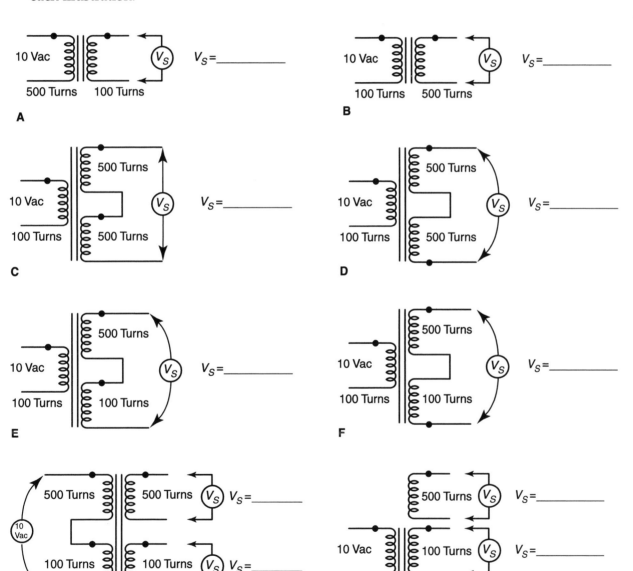

Figure 30-1.

2. Using the transformer connections shown in Figure 30-1, calculate the indicated secondary voltages of each combination. Be sure to observe the polarity dot which indicates the beginning of the winding.

3. If multiple-winding transformers such as those shown in Figure 30-1 are available in the laboratory, perform secondary voltage measurements with different winding combinations and a constant voltage input.

Analysis

Briefly discuss turns ratio, voltage ratio, and current ratio of single-phase transformers.

_____

_____

_____

_____

_____

Activity 31–Transformer Problems

Name _____ Date _____ Score _____

Procedure

Solve the following problems dealing with transformer operation. Record all your answers on a separate sheet of paper. Show all of your work.

1. A 1500-turn primary winding has 120 V applied. The secondary voltage is 20 V. How many turns of wire does the secondary winding have?

2. When a 200-Ω resistance is placed across a 240-V secondary winding, the primary current is 30 A. What is the value of the primary voltage of the transformer?

3. A transformer has a 20:1 turns ratio. The primary voltage is 4800 V. What is the secondary voltage?

4. A transformer has a 2400-V primary winding and a 120-V secondary. The primary winding has 1000 turns. How many turns of wire does the secondary winding have?

5. A 240- to 4800-V transformer has a primary current of 95 A and a secondary current of 4 A. What is its efficiency?

6. A single-phase transformer has a power rating of 20 kVA. Its primary voltage is 120 V and its secondary voltage is 480 V. What are the values of its maximum primary current and secondary current?

Activity 32–Autotransformer Principles

Name _____ Date _____ Score _____

Objectives

The autotransformer has only one winding. Therefore, it requires less copper wire to construct than a conventional single-phase transformer of similar size. The major disadvantage of an autotransformer is the necessity of having a direct connection to one of the primary input lines.

In this laboratory activity, you will observe the operational principles of an autotransformer. You can use either an autotransformer or a secondary winding of a center-tapped transformer. You should notice that this device can be constructed as a step-up or step-down transformer.

Equipment and Materials

- Multimeter
- Variable ac power source
- Center-tapped transformer
- Resistors—1000 Ω, 10,000 Ω, 10 W
- AC current meter

Procedure

1. Construct the autotransformer circuit shown in Figure 32-1.

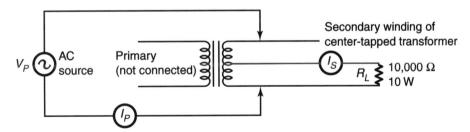

Figure 32-1. Autotransformer step-down connections.

2. Apply rated voltage to the outer winding terminals of the center-tapped transformer. Measure the voltage across the load (V_{R_L}), primary current (I_P), and secondary current (I_S).

V_{R_L} = _____ volts ac.

I_P = _____ amperes ac.

I_S = _____ amperes ac.

3. Change the value of load resistance to 1000 ohms and repeat the voltage and current measurements.

V_{R_L} = _____ volts ac.

I_P = _____ amperes ac.

I_S = _____ amperes ac.

4. Modify the circuit as shown in Figure 32-2. The ac input voltage should remain the same as before.

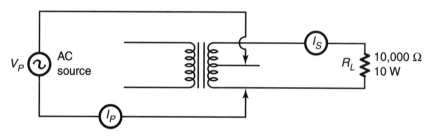

Figure 32-2. Autotransformer step-up connections.

5. Again, measure V_{R_L}, I_P, and I_S.

V_{R_L} = _____ volts ac.

I_P = _____ amperes ac.

I_S = _____ amperes ac.

6. Change the value of load resistance to 1000 ohms and repeat the measurements.

V_{R_L} = _____ volts ac.

I_P = _____ amperes ac.

I_S = _____ amperes ac.

Analysis

1. What type of autotransformer was constructed in Step 1? _____

   _____

2. What type of autotransformer was constructed in Step 4? _____

   _____

3. From Step 2, verify that $V_P \times I_P \approx V_S \times I_S$. _____

4. From Step 3, verify that $V_P \times I_P \approx V_S \times I_S$. _____

5. From Step 5, verify that $V_P \times I_P \approx V_S \times I_S$. _____

6. From Step 6, verify that $V_P \times I_P \approx V_S \times I_S$. _____

7. What are some applications of autotransformers? _____

   _____

8. How does decreased load resistance affect the primary current? _____

   _____

9. Are the primary and secondary power values ($V_P \times I_P$ and $V_S \times I_S$) equal? Why? _____

   _____

   _____

Activity 33—Electrical Conductors and Insulators

Name _____ Date _____ Score _____

Objectives

Two very important parts of industrial power distribution systems are conductors and insulators. Conductors of various types are designed to carry electrical current. Conductors can be in the form of solid wires, stranded cables, cords, bus bars, or many other types. The amount of current that a conductor can safely handle is mainly determined by its diameter. Examples of several conductor tables are included in Chapter 6 in your textbook.

The diameter of a conductor is measured with an American Wire Gage (AWG). The sizes from No. 18 through 1000 MCM (1000 circular mils) are shown in Figure 6-17 of your textbook.

The insulation used on electrical conductors is mainly determined by the type of location where it will be used and the type of enclosure or wireway where it will be placed. The symbols R, H, W, T, and N indicate rubber, heat-resistance, moisture-resistance, thermoplastic, and nylon, respectively. A table of some insulation types is shown in Figure 6-24 of your textbook.

In this activity, you will observe several types of conductors and insulation. You should use samples of conductors provided by your instructor.

Equipment and Materials

- American Wire Gage (AWG)
- Conductor samples

Procedure

1. Obtain an American Wire Gage and several conductors from your instructor.

2. Measure the diameter of each conductor with the AWG and record the results in the following spaces. Also record whether each conductor is copper or aluminum.

   Sizes of conductors:

   a. _____

   b. _____

   c. _____

   d. _____

   e. _____

   f. _____

3. Refer to Figures 6-20 and 6-21 in your textbook to determine the amount of current each conductor can carry. Record your answers in the following spaces.

Ampacity of conductors:

a. _____ amperes (raceway) _____ amperes (free air)

b. _____ amperes (raceway) _____ amperes (free air)

c. _____ amperes (raceway) _____ amperes (free air)

d. _____ amperes (raceway) _____ amperes (free air)

e. _____ amperes (raceway) _____ amperes (free air)

f. _____ amperes (raceway) _____ amperes (free air)

4. List the type of insulation on each conductor and record in the following spaces.

Type of insulation:

a. _____

b. _____

c. _____

d. _____

e. _____

f. _____

Analysis

1. What is the area in circular mils of the following conductors? (Use Figure 6-17 from your textbook.)

a. No. 16 = _____ cmil.

b. No. 0 = _____ cmil.

c. No. 8 = _____ cmil.

d. 250 MCM = _____ cmil.

e. 1000 MCM = _____ cmil.

f. No. 12 = _____ cmil.

2. How many wires are used to construct the conductors listed in the preceding question?

a. _____

b. _____

c. _____

d. _____

e. _____

f. _____

3. What are the types of the following insulations?

   a. RH = _____

   b. RHW = _____

   c. THW = _____

4. What are the ampacities of the following conductors in a raceway?

   a. No. 12 (RH) copper = _____ A.

   b. No. 4 (TW) aluminum = _____ A.

   c. 250 MCM (THW) aluminum = _____ A.

   d. No. 6 (T) copper = _____ A.

   e. No. 000 (R) aluminum = _____ A.

   f. 1000 MCM copper = _____ A.

5. What is the dc resistance of the following lengths of conductors?

   a. 500 feet of No. 8 copper = _____ Ω.

   b. 800 feet of No. 12 aluminum = _____ Ω.

   c. 5000 feet of 250 MCM aluminum = _____ Ω.

   d. 1500 feet of 500 MCM copper = _____ Ω.

   e. 400 feet of No. 6 copper = _____ Ω.

   f. 8000 feet of 1000 MCM aluminum = _____ Ω.

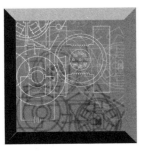

Activity 34–National Electrical Code (Self Test)

Name _____ Date _____ Score _____

Objectives

The *National Electrical Code (NEC)* is a very important document to understand. All industrial equipment and wiring must conform to the NEC standards. The NEC is not difficult to use. The user should become familiar with the comprehensive index contained in the NEC and the organization of the various sections. For instance, if you wished to review the standards related to system grounding, you should look in the index and locate this term. The index will refer you to the appropriate sections in the NEC that discuss system grounding.

In this activity, you will need to review the latest edition of the NEC. Once you have learned how to use it, you should complete the self test on the NEC.

Self Test

Using a recent copy of the NEC, find the correct responses to complete each of the statements below. Place the article number/page number where the answer was obtained in the parentheses at the end.

1. A new edition of the NEC is published every _____ years. (_____)

2. A No. 8 copper conductor in free air with THW insulation is capable of conducting _____ amperes of current. (_____)

3. When selecting nonmetallic-sheathed cable for general use, you should specify type _____. (_____)

4. If you were wiring for an air conditioner, you should refer to Article _____ in the NEC. (_____)

5. The diameter of electrical conductors is measured with a(n) _____ gage. (_____)

6. Line-to-ground shock hazards can be reduced by utilizing _____ _____ _____. (_____)

7. The definition of a "ground" can be located in Article _____ of the NEC. (_____)

8. Article 402 of the NEC specifies that type CF fixture wire has _____ insulation. (_____)

9. If you were installing a three-phase motor with associated control circuits, you should refer to Article _____ of the NEC. (_____)

10. Article 810 of the NEC provides electrical wiring specifications for _____. (_____)

11. Underground feeder cable comes in sizes _____ to _____ inclusive. (_____)

12. The full-load current of a one-horsepower, 120-volt dc motor as listed in the NEC is _____ amps. (_____)

13. The full-load current of a 30-horsepower, three-phase ac induction motor operating on 208 volts is _____ amps. (See information at bottom of the table for conversion.) (_____)

14. According to Article 210 of the NEC, grounded conductors in a branch circuit should be color-coded _____ or _____. (_____)

15. The general lighting load of an office building is specified to be _____ watts per square foot of floor space. (_____)

Activity 35–Power Distribution Substation Visit

Name _____ Date _____ Score _____

Objectives

After completing this laboratory activity, you will be able to identify and explain the function of equipment used in electrical power substations.

Procedure

Visit any distribution substation. You should be able to observe the construction and operation of the following equipment. Take notes describing what you see.

1. High voltage conductors.

2. Wood poles and crossarms.

3. High voltage insulators.

4. High voltage transformers.

5. Voltage regulators.

6. Oil-filled circuit breakers.

7. Bus systems.

8. Lightning arresters.

9. Air break switches.

10. Metal-clad switchboards.

Analysis

Discuss each of the items of equipment you observed.

Activity 36–Distribution for Electrical Lighting

Name _____ Date _____ Score _____

Objectives

In the laboratory activity at the end of Chapter 7 in your textbook, you learned how to wire a power distribution panel. This activity will illustrate the proper methods for wiring electrical lighting circuits for different areas of a building. Some lighting fixtures are controlled from one point by one switch. Other lighting fixtures can be controlled from two or more points by a switch at each point.

Electrical lighting circuits are typical branch circuits. The most common type of lighting circuit is a 120-volt branch, which extends from the power distribution panel to the light fixture or fixtures of some area of a building. The path for electrical distribution is controlled by one or more switches, which are usually placed in small metal enclosures inside a wall and covered by rectangular plastic covers.

Switches are always placed in a wiring circuit so that they open or close a *hot* wire that distributes electrical power to the lighting fixtures that they control. A switch that accomplishes control from one location is a simple single-pole, single-throw (SPST) switch. Control from two locations, such as an outside door and an inside door of a home, is accomplished by two three-way switches. When control of a lighting fixture from more than two points is desired, two three-way switches and one or more four-way switches are used. For instance, control of one light from five points could be accomplished by using a combination of two three-way switches and three four-way switches. The three-way switches are always connected to the power panel and to the light fixture, with the four-way switches between them. Using three-way and four-way switch combinations, it is possible to achieve control of a light from any number of points.

In this activity, you will wire circuits that achieve control of one light from one, two, and three points, respectively. Each involves a different type of switching combination to accomplish control of a lighting fixture. The same type of circuits are commonly wired in the home to provide the types of lighting desired for each room.

In this laboratory activity, you will also wire a light dimmer circuit for controlling an incandescent light. Light dimmers are relatively simple to install and their installation is commonly done in the home to control dining room lighting.

A light dimmer is ordinarily placed in a lighting circuit as a substitute for a switch that turns a light on and off. The light dimmer provides a wide range of light intensity in a room by changing the adjustment of a potentiometer that is mounted into the wall. This light dimmer takes the place of a switch that does not provide the convenience of light intensity variation.

Equipment and Materials

- Multimeter
- Single-phase, three-wire ac power source
- Switchboxes (3)
- Single-pole, single-throw (SPST) switch
- Three-way switches (2)
- Four-way switch
- Power distribution panel
- Octagonal box
- Porcelain lamp receptacle
- Light bulb—60-watt
- Box connectors
- Solderless connectors
- No. 12-2 WG wire (4 pieces cut to desired lengths)
- No. 12-3 WG wire (2 pieces)
- Wire strippers
- Cable-ripping tool
- Screwdriver
- Needle-nose pliers
- Light dimmer

Safety

Be very careful when working with high voltages.

Procedure

Section A: Light Controlled from One Location

1. Use the power distribution panel you produced in the laboratory activity at the end of Chapter 7, which was wired to deliver power to several branch circuits. Also, obtain three metal switchboxes and a metal hexagonal holder for a porcelain lamp receptacle. These may be mounted on a board.

2. Prepare the distribution panel for operation as outlined in the activity from Chapter 7.

3. Run a 120-volt branch circuit using a No. 12-2 WG cable from one of the plug fuse terminals and the neutral/ground terminal strip to a metal switch box. Both ends of the cable should be secured by box connectors. About 1/2 inch of insulation should be stripped from the ends of each conductor before connecting these to the terminal strips.

4. Leave sufficient lengths of wire extending from the box connectors to allow easy connection to the terminal screws. Remember that the hot (black) wire should be connected to the fuse terminal and the neutral (white) wire and the safety ground (bare or green) should be connected to the neutral/ground terminal strip. Refer to Figure 36-1 to wire a branch circuit that controls a light from one location.

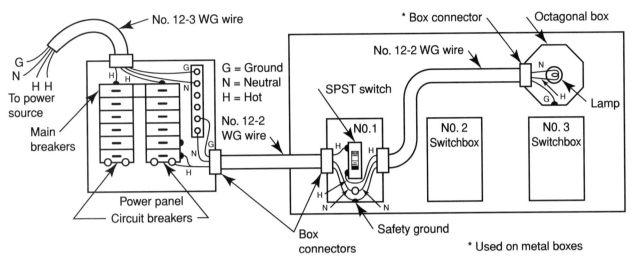

Figure 36-1. Wiring of branch circuit to light.

5. Obtain a single-pole, single-throw (SPST) switch for mounting in a metal switch box. Use an ohmmeter (set on lowest resistance range) to check across the switch terminals. What occurs when the switch is turned on and off? _____

_____

6. Connect a No. 12-2 WG cable from the switch box to the lamp receptacle box. Use box connectors at both ends. Prepare the wire as outlined in step 3.

7. Place the SPST switch above the switchbox. Connect the hot wire to one terminal of the switch.

8. Place the other hot wire in the switchbox onto the SPST switch terminal. The neutral wires inside the switchbox should be secured together by using a solderless connector. The safety ground conductors should be connected directly to the metal switchbox. Tighten the screws which fasten the switch onto the switchbox.

9. Using a porcelain light receptacle with a 60-watt light bulb, prepare the light fixture for operation. The hot wire should be connected to the darker colored screw terminal and the neutral wire to the lighter colored screw terminal. The safety ground conductor should be connected directly to the octagonal box. Do not tighten the screws that fasten the porcelain receptacle onto the hexagonal box yet.

10. Before connecting the input of the power distribution panel to the power source, connect an ohmmeter across the light terminals. Turn the switch on and off. What occurs? _____

_____

11. Now secure the lamp receptacle with screws and connect the power source to the distribution panel.

12. Turn on the power source and check the circuit for operation.

13. Have your instructor check your circuit for correct operation and wiring methods.

Instructor's approval _____

14. Turn off the power source.

Activity 36—Distribution for Electrical Lighting

Section B: Light Controlled from Two Locations

1. Disconnect the SPST switch from the switch box.

2. Move the No. 12-2 WG cable from the right side of the switchbox at location 1 on the previous diagram to the right side of the switchbox at location 2.

3. Run a No. 12-3 WG cable from the right side of switchbox 1 to the left side of switchbox 2. The wires between the switches are called *travelers*.

4. Obtain two three-way switches. Use an ohmmeter to locate the hinge points of these switches. The hinge point is illustrated in Figure 36-2. Be sure to check properly for determining the hinge points of each three-way switch.

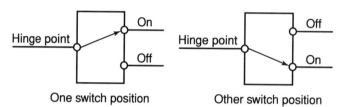

Figure 36-2. Switch positions of three-way switch.

5. Place the three-way switches above the switchboxes and connect the terminal screws as shown pictorially in Figure 36-3. Observe the position of the hinge points.

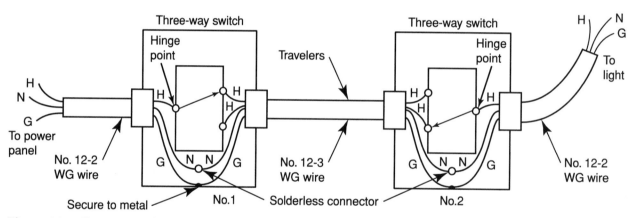

Figure 36-3. Connecting three-way switches.

6. Secure the switches onto the switches boxes by using screws.

7. Turn on the power source and check the circuit for operation.

8. Have your instructor check your circuit for correct operation and wiring methods.

   Instructor's approval _____

9. Turn off the power source.

Section C: Light Controlled from Three Locations

1. Disconnect the three-way switch at position 2.

2. Move the cable from the light to the right of switchbox 2.

3. Obtain a four-way switch. Use an ohmmeter to check the internal continuity of the switch. The four terminals are interconnected as shown in Figure 36-4. Verify these connections using the ohmmeter.

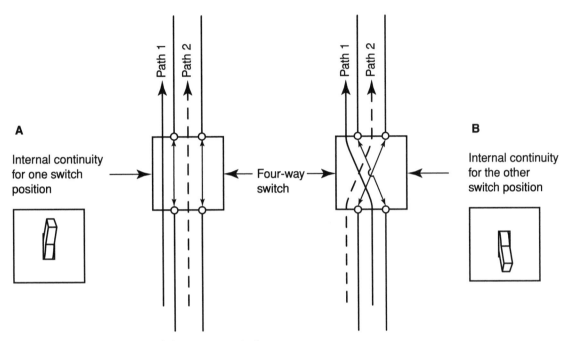

Figure 36-4. Wiring of three- and four-way switches.

4. Wire the switch combination shown pictorially in Figure 36-5.

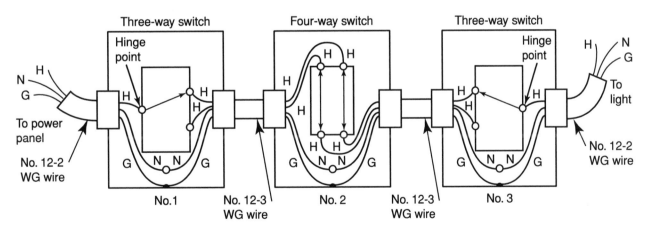

Figure 36-5. Wiring a switch combination.

5. Secure the switches onto the switch boxes with screws.

6. Turn on the power source and check the circuit for operation.

7. Have your instructor check your circuit for correct operation and wiring methods.

   Instructor's approval _____

8. Turn off the power source and disconnect the switches and the light. Leave the wires connected to the power source.

Section D: Light Dimmer Circuit

1. Wire the power distribution panel to supply a 120-volt branch circuit as shown in Figure 36-6. Connect a No. 12-2 WG branch circuit into one of the switch boxes on the lighting circuit board.

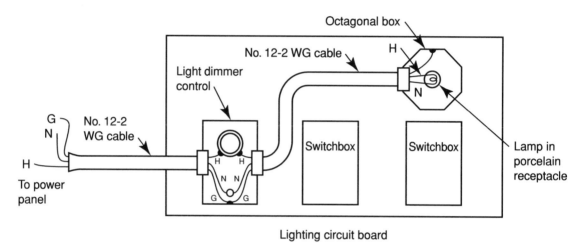

Figure 36-6. Wiring of light dimmer circuit.

2. Run a No. 12-2 WG wire from the switchbox to the lamp box.

3. Place the light dimmer near the switchbox and connect it in series with the hot lines (black wires) of the branch circuit.

4. Connect the neutrals together using a solderless connector.

5. Connect the ground wires together and secure them to the switchbox. Fasten the dimmer onto the switchbox.

6. Connect the No. 12-2 WG cable from the switchbox to the porcelain light receptacle and attach the hot and neutral wires of the cable to the terminals of the receptacle.

7. Attach the receptacle onto the octagonal box.

8. Connect the No. 12-3 WG wire from the power source to the distribution panel.

9. Turn on the power source and check the light dimmer for operation.

10. Have your instructor check your circuit for correct operation and wiring methods.

    Instructor's approval _____

11. Disassemble the light dimmer circuit and return all equipment to the proper storage areas.

Analysis

1. What size of a conductor should be used for lighting circuits in the home?_____

2. Make a simplified sketch showing how a light could be controlled from four points.

3. Why must the hot and neutral wires connected to the lamp receptacle be placed on the proper terminals? _____
   _____

4. What is the purpose of the safety ground in lighting circuits? _____
   _____

5. How many and what type of switches would be required to control a light from 10 points?
   _____

6. Explain how a light dimmer is connected into a branch circuit. _____
   _____

7. What determines the power rating requirements of a light dimmer?_____
   _____

8. What are some reasons for using a light dimmer to control light intensity?_____
   _____

9. Could the light dimmer used in this activity be used with a light controlled from more than one point? Why? _____
   _____

Activity 37–Distribution for Duplex Outlets and Door Chimes

Name _____ Date _____ Score _____

Objectives

In this activity, you will learn the proper method of wiring two common types of branch circuits. One of the most used branch circuits is for duplex power outlets. The other type of branch circuit you will wire is a door chime circuit.

Duplex power outlets are connected in parallel throughout the various rooms of a building. When a small appliance or room lamp is plugged into the duplex outlet, 120 volts is applied to the appliance or lamp. The branch circuit for the duplex outlets begins at the power distribution panel and extends to the first duplex outlet. The number of duplex outlets that can be connected in parallel with the first outlet is determined by the power requirements of the room or area of the building. The number of outlets is also limited by the current rating of the circuit breaker used to protect the branch circuit.

Door chimes are used in many homes and commercial buildings. The branch circuit used to deliver electrical power to door chimes is a special type of circuit. This branch circuit uses a step-down transformer to reduce the 120 volts from the power distribution panel to a lower voltage, such as 12 volts. The 12 volts is used to energize the solenoids or electromagnetic coils of the door chimes. Two momentary-on push-button switches are used in this activity to represent the push buttons at the front door and back door of a building. One push button energizes one of the solenoids, which represents the front door sound, and the other energizes the second solenoid, which represents the back door sound.

Equipment and Materials

- Multimeter
- Single-phase, three-wire power source
- Neon circuit tester
- Duplex power outlets (2)
- Outlet boxes (2)
- Door chimes
- Push-button switches—momentary-on (2)
- Power distribution panel
- Transformer—120-volt primary and 12-volt secondary
- Octagonal box
- No. 12-2 WG nmc cable (3 pieces)
- No. 26 single-strand wire (for door chimes)
- Screwdriver

- Wire strippers
- Knife
- Cable-ripping tool
- Box connectors
- Solderless connectors
- Needle-nose pliers

Safety

Be very careful when working with high voltages.

Procedure

1. Refer to Figure 37-1 to wire the branch circuits for duplex power outlets and door chimes.

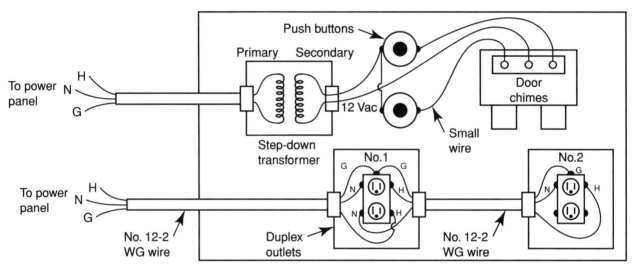

Figure 37-1. Wiring of branch circuits.

2. Run a No. 12-2 WG cable from the power distribution panel through a box connector into the metal outlet box of the No. 1 duplex outlet. Ensure that the hot wire connects through a breaker and the neutral wire is grounded in the power panel.

3. Run a piece of No. 12-2 WG cable from outlet box 1 to outlet box 2. Use box connectors for each end of the cable.

4. Place a duplex outlet above the outlet box and connect the No. 12-2 WG cable to the screw terminals of the outlet. The hot wires should *always* be connected to the darker colored terminal and the neutral wires to the lighter colored terminal. The safety ground wire connects to the green colored terminal. These precautions are very important for *safety* reasons.

5. Connect the wires of the other No. 12-2 cable to the terminals of duplex power outlet 1. Use the same procedure to secure the outlet onto the box by using screws.

6. Now connect the wires of outlet box 2 to a duplex outlet using the same procedure. Secure this outlet to the outlet box. This completes the duplex power outlet circuit wiring.

7. Run a No. 12-2 WG cable from the power distribution panel to the primary winding of the step-down transformer. Use solderless terminals to connect the cable to the transformer wires. This branch circuit will energize the door chimes.

8. Use the small wire to connect the door chimes and push-button switches to the low-voltage secondary winding terminals of the transformer. This will apply 12 volts to the door chimes when the circuit is energized.

9. Connect the input of the power distribution panel to the single-phase, three-wire source.

10. Turn on the power source. Be very careful.

11. Test the duplex power outlets with a neon circuit tester.

12. Test the door chimes by pressing the push-button switches.

13. Have your instructor check your circuits for operation and wiring methods.

    Instructor's approval: _____

14. Turn off the power source and disassemble the branch circuits.

Analysis

1. Why are door chimes controlled by a low voltage? _____

   _____

   _____

2. Why must the hot wires of duplex outlet circuits be connected to the darker colored terminals of the outlet? _____

   _____

3. Explain the method used to identify electrical wiring cable, for example No. 10-2 WG nmc.

   _____

   _____

4. Are duplex outlets connected in series or parallel? Why? _____

   _____

Activity 38–Three-Phase Transformer Schematics

Name _____ Date _____ Score _____

Objectives

Upon completion of this activity, you will be familiar with connections of five common types of transformer connections.

Procedure

Complete each of the following diagrams.

1. Delta–delta connection.

Primary

A _____

B _____

C _____

N _____

N _____

A _____

Secondary

B _____

C _____

2. Delta–wye connection.

3. Wye–wye connection.

4. Wye–delta connection.

Primary

C _____

B _____

A _____

N _____

Secondary

N _____

A _____

B _____

C _____

5. Open delta connection.

Primary

C _____

B _____

A _____

Shorted
secondary

Secondary

A _____

B _____

C _____

Activity 39–Three-Phase Transformer Calculations

Name _____ Date _____ Score _____

Problems

Given:

- A three-phase transformer connected in a *delta-wye* configuration.

- A 2:1 turns ratio.

- Primary line voltage—V_L (Pri) = 600 V.

- Primary phase voltage—V_P (Pri) = 600 V.

- Primary line current—I_L (Pri) = 200 A.

Calculate:

1. I_P (Pri) = _____ A.

2. V_L (Sec) = _____ V.

3. V_P (Sec) = _____ V.

4. I_L (Sec) = _____ A.

5. I_P (Sec) = _____ A.

6. P_P (Sec) = _____ kW.

7. P_T (Sec) = _____ kW.

Activity 40–Electric Motor Principles

Name _____ Date _____ Score _____

Objectives

In this activity, you will study the operation of an electric motor. Motors convert electrical energy into mechanical energy in the form of rotary motion. The type of motor you will study is called a universal motor. It will operate with either alternating current or direct current applied.

The universal motor is one of the most used motors. Applications of the universal motor include electric drills, electric saws, and household items such as mixers, blenders, and sewing machines. The universal motor is one of the few types of variable-speed motors that operates on alternating current. Most types of ac motors operate at a constant speed. Thus the primary advantage of a universal motor is its speed control capability. By varying the applied voltage, the speed of the motor will change.

The universal motor has several basic parts. One part is the rotor, or armature, which is constructed of many windings of copper wire and which is embedded in a laminated metal core assembly. A second part is the stator, or field, which is the metal enclosure and at least one set of field coils which are mounted on laminated metal pole pieces. The armature windings are connected to the power source and the field windings through a brush/commutator assembly. The brushes make contact with the rotating commutator segments to apply electrical power to the armature. The commutator segments, which are made of copper, provide a conductive path through the armature windings. The field coils are also constructed of copper windings that develop a concentrated magnetic field since they are actually electromagnets. The interaction of the electromagnetic fields of the field windings and the armature windings causes a force to be developed that will produce rotation.

In this activity, you will use dc voltage to energize the universal motor. Keep in mind that this type of motor will also operate with alternating current applied to it. You will examine several of the important characteristics of the universal motor in the following procedure.

Equipment and Materials

- Multimeter
- Universal motor
- Variable dc power supply

Safety

Be sure to wear eye protection while electric motors are in operation. Also, be very careful when working with high voltages.

Procedure

1. The connection diagram for a universal motor is shown in Figure 40-1. Note that the armature and field coils are connected in series with the power.

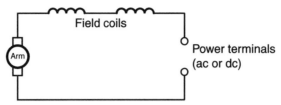

Figure 40-1. Universal motor schematic diagram.

2. Connect the power terminals of the universal motor to the variable dc power supply. The voltage should be set at zero.

3. Turn on the power supply and make the necessary measurements to complete Figure 40-2. The current should be measured with the meter in series with either one of the power terminals.

**Motor Voltage, Current, and
Speed Characteristics**

Applied voltage	Current	Relative speed
10		
20		
30		
40		

Figure 40-2. Complete the chart.

4. Set the power supply to 30 volts dc. Then turn the motor off. Keep the ammeter in series with the motor terminals.

5. Turn the motor on and record the initial current (starting current) and the running current.

   Starting current = _____ A; Running current = _____A.

6. While the motor is running on 30 Vdc, carefully apply a load to its shaft. This can be done by grasping the shaft very *carefully* with your hand. What is the effect on current as the load is increased? _____

   _____

   _____

7. Look toward the external shaft of the motor and observe the direction of rotation.

   The direction of rotation is: clockwise or counterclockwise (circle one).

8. Reverse the motor terminal connections to the power supply.

   Does the direction of rotation change? _____

Analysis

1. Complete the following statements, which refer to the operation of a universal motor, by using the words *increases* or *decreases*.

   a. As applied voltage increases, current _____.

   b. As applied voltage increases, speed _____.

   c. As load applied to the shaft of the motor increases, speed _____.

2. In Step 5, why were the starting and running currents different?

   _____

   _____

   _____

3. How can the direction of rotation of a universal motor be reversed?

   _____

   _____

   _____

4. What are the basic parts of a universal motor and what is the function of each part?

   _____

   _____

   _____

Activity 41—Series-Wound DC Motors

Name _____ Date _____ Score _____

Objectives

The series-wound dc motor is constructed of low-resistance field coils connected in series with the armature coils. The field coils have a few turns of large diameter wire. This arrangement permits a current flow that is the same throughout the circuit. Like all motors, the series-wound machine slows down with increased loads and speeds up with decreased loads. Its speed, like other dc motors, is inversely proportional to its magnetic field strength. The counterelectromotive force, or cemf, which is generated when the armature turns through the magnetic field, is directly related to speed. As a result, these three factors, speed, cemf, and field strength, are extremely important characteristics in the operation of a series-wound dc motor.

In this activity, you will analyze the operation of a series-wound dc motor. This type of motor is commonly designed so that it will operate from either a dc or ac voltage source. Such machines are called universal motors. This activity is set up in such a way that you can analyze: (1) a commercially manufactured dc series motor, such as a small universal motor used in an electric drill or an automobile starter, or (2) a dc series motor that has been constructed using a rotating machinery unit.

Equipment and Materials

- Multimeter
- Series-wound dc motor
- DC power supply
- Tachometer

Safety

Be sure to wear eye protection while electric motors are in operation. Also, be very careful when working with high voltages.

Procedure

1. Examine the connection diagram for a series-wound dc motor, shown in Figure 41-1.

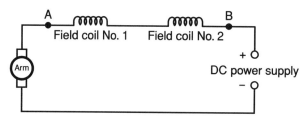

Figure 41-1. Connection diagram for a series-wound dc motor.

2. Obtain a dc series motor or construct one using the rotating machinery unit.

3. Measure the resistance of the armature. This will be a low resistance, so use the lowest resistance range of the meter.

   $R_A =$ _____ W.

4. Measure the resistance of the field coils (point A to point B on the diagram). Make sure the coils are isolated from other parts of the circuit. It may be necessary to remove a wire at point A.

   $R_F =$ _____ Ω.

5. Prepare the multimeter to measure direct current and connect it in series with the motor circuit.

6. Gradually increase the dc voltage applied to the machine until you reach its rated voltage. Observe the effect on current and speed.

7. With rated voltage applied, turn the power supply off and then back on momentarily. Observe the effect on current as speed increases toward its maximum.

8. Carefully apply a load on the motor shaft while observing the effect on current and speed.

9. With a tachometer, measure the motor speed at its rated voltage.

   Speed = _____ r/min.

10. Record the current indicated at the rated voltage of the machine.

    Current = _____ A.

11. Turn the power supply off and use the necessary procedure to *reverse* the direction of rotation of the motor.

12. Turn on the power supply and measure the speed and current of this machine with its direction of rotation reversed.

    Speed = _____ r/min.

    Current = _____ A.

13. Try to apply a heavy enough load to stall the motor. Very quickly record the maximum current the motor draws.

    Maximum current = _____ A.

14. Allow the motor to run for a few seconds to cool off. Then, turn off the power and return all materials.

Analysis

1. What effect would having more than two field coils have on the operation of this motor? ____

_____

2. What effect does increased voltage have on motor speed? Why? (See Step 6). _____

_____

3. In Step 7, what happened to motor speed and current as load was increased? Why? _____

_____

4. How is the direction of rotation of the dc series motor reversed? _____

_____

5. Was there any change of motor speed or current when the direction of rotation was reversed? Should there have been? _____

_____

6. Show in the graph of Figure 41-2, how a speed-versus-load curve should look for a series dc motor.

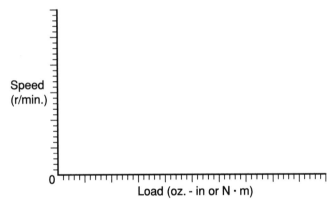

Figure 41-2. Graph of speed versus load for a series-wound dc motor.

7. Calculate the input power to this motor at its rated voltage.

Watts input = $V \times I$ = _____ W.

8. Convert the watts (W) input to horsepower (hp) using the following formula.

hp = W/746 = _____.

Note that this calculated value is not the output horsepower rating of the machine. The output horsepower would be somewhat lower due to energy losses of the machine.

Activity 42–Shunt-Wound DC Motors

Name _____ Date _____ Score _____

Objectives

The construction of dc motors is based on the method of connecting the field windings and armature coils together. In the shunt-wound dc motor, for example, the field coils are connected in parallel with the armature. This is achieved by making the shunt field a fairly high resistance. Accordingly, it is wound of many turns of fine wire. The armature, however, offers less resistance and permits a larger current to flow. Its construction is of a larger wire with fewer turns of wire.

In the laboratory activity, you will analyze either: (1) a commercially manufactured shunt-wound dc motor, or (2) a shunt-wound dc motor that has been constructed in the laboratory using a rotating machinery unit. You should observe the general construction and operational characteristics of the shunt-wound dc motor.

Equipment and Materials

- Multimeter
- Shunt-wound dc motor
- Variable dc power supply
- Field rheostat (0-500 Ω, high wattage)
- Armature rheostat (0-25 Ω, high wattage)
- Tachometer

Safety

Be sure to wear eye protection while electric motors are in operation. Also, be very careful when working with high voltages.

Procedure

Section A: Shunt-Wound DC Motor Characteristics

In this section of the laboratory activity, you will observe the operational characteristics of a shunt-wound dc motor. You will observe the effect of load on the speed, counterelectromotive force (cemf), current, and torque of this machine.

1. The connection diagram for a shunt-wound dc motor is shown in Figure 42-1. Set up this circuit.

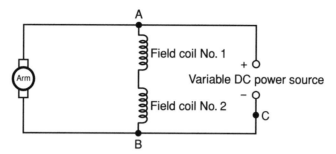

Figure 42-1. Connection diagram for a shunt-wound dc motor.

2. Measure the resistance of the field coils (point A to point B on the diagram). You will have to remove a wire at point A. It will be necessary to use the lowest resistance range on your meter.

   $R_F =$ _____ Ω.

3. Measure the resistance of the armature circuit (across the brushes). Make sure the power supply is not connected to the armature, and the wire at point A is still removed. This will isolate the armature resistance from other circuits.

   $R_A =$ _____ Ω.

4. Prepare a meter to measure a high value of dc current. Put the current meter in series with the power supply (at point C on the diagram). This meter will measure the total current drawn by the motor.

5. Record the rated voltage value of this motor.

   Rated voltage = _____ volts dc.

6. Turn on the power supply and slowly increase the voltage to the rated voltage of the motor. Observe the effect on current and speed of the motor.

7. Record the total current (I_T) drawn by the motor at its rated voltage.

   $I_T =$ _____ amperes dc.

8. With a tachometer, measure and record the speed of the motor.

   Speed = _____ r/min.

9. Carefully apply a load to the motor shaft by putting pressure on the shaft, which will slow its speed. Observe the effect that increased load has on speed and total current.

10. Increase the load as much as possible, or until rotation stops or slows down appreciably. Quickly record the maximum current drawn by the motor.

    Maximum current = _____ amperes dc.

11. Turn off the power supply and perform the necessary procedure to reverse the direction of rotation of the motor.

12. With the direction of rotation reversed, increase the applied dc voltage to the rated voltage value of the motor.

Activity 42—Shunt-Wound DC Motors

13. Measure and record the current and speed.

   Current = _____ amperes dc.

   Speed = _____ r/min.

14. Turn off the power supply.

Section B: Speed Control of Shunt-Wound DC Motors

   In this section, you will observe two methods of controlling the speed of a dc shunt-wound motor. These methods are very simple. The first method utilizes a rheostat in series with the armature circuit. The second method employs a rheostat in series with the field circuit. By measuring current through each rheostat and calculating power dissipation, you should be able to decide which speed control method is most desirable.

1. Modify the dc shunt motor as shown in Figure 42-2. Be sure the wattage ratings of the rheostats are high enough.

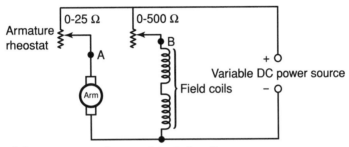

Figure 42-2. Shunt-wound dc motor speed-control test circuit.

2. The rheostats should be connected in series with the armature and field circuit as shown in the diagram. Adjust each rheostat to the zero resistance position.

3. Adjust the variable dc power supply from zero volts to the rated voltage of the motor to check it for operation.

4. Turn off the power supply and connect a dc current meter in series with the armature circuit at point A on the diagram.

5. With rated voltage applied, turn on the motor and observe the maximum starting current indicated on the meter.

   Maximum starting current = _____ amperes dc.

6. Measure and record the running current of the motor with no load applied.

   Running current = _____ amperes dc.

7. Adjust the armature rheostat from zero toward its maximum resistance while observing the speed of the motor. As armature resistance increases, speed _____. With a tachometer, measure the range of speed.

   Speed range = _____ to _____ r/min.

8. Adjust the armature rheostat back to zero resistance. Carefully apply a load to the motor shaft. Measure and record the maximum armature current as load is increased.

   Maximum armature current = _____ amperes dc.

9. Turn off the power supply. Now, remove the armature current meter and connect the meter in series with the field circuit at point B on the diagram.

Activity 42—Shunt-Wound DC Motors

10. With rated voltage applied, turn on the motor and observe the maximum field current during starting.

   Maximum field current = _____ amperes dc.

11. Record the field current indicated while the motor is running with no load applied.

   Field current = _____ amperes dc.

12. Adjust the field rheostat from zero toward its maximum resistance while observing the speed of the motor. As field resistance increases, speed _____. With a tachometer, measure the range of speed.

   Speed range = _____ to _____ r/min.

13. Adjust the field rheostat back to zero resistance. Carefully apply a load to the shaft of the motor. Measure and record the maximum field current as load is increased.

   Maximum field current = _____ amperes dc.

14. Turn off the power supply. Return all materials.

Analysis

1. How does the field resistance of a shunt-wound dc motor compare to that of a dc series motor?

   _____

   _____

   _____

2. How does increased load on a shunt-wound dc motor affect:

   a. Speed of the motor? _____

   _____

   b. Cemf developed by the armature? _____

   _____

   c. Total current drawn by the motor? _____

   _____

   d. Torque developed by the motor? _____

   _____

3. In Section A, Step 11, how was the direction of rotation of the motor reversed? _____

   _____

   _____

   _____

4. Did reversing the direction of rotation have any effect on total current and speed? (See Section A, Steps 7, 8, and 13.) Why?

   _____

   _____

   _____

5. How do the starting current and running current drawn by the armature compare? (See Section B, Steps 5 and 6.) Why? _____

_____

_____

6. Using the maximum armature current value from Section B, Step 5, and the measured armature resistance, determine the minimum power rating of an armature rheostat.

$P_A = I_A{}^2 \times R_A = $ _____ W.

7. In Section B, Step 7, as armature resistance changes, why does speed change as it does? _____

_____

_____

8. How do starting current and running current drawn by the field compare? (See Section B, Steps 10 and 11.) Why? _____

_____

_____

9. In Section B, Step 12, as field resistance changes, why does speed change as it does? _____

_____

_____

_____

10. Using the maximum field current obtained in Section B, Step 13, and the measured field resistance, determine the minimum power rating of a field rheostat for speed control.

$P_F = I_F{}^2 \times R_F = $ _____ W.

11. Which type of speed control (armature or field) is most desirable? Why? _____

_____

_____

_____

Activity 43–Single-Phase AC Capacitor-Start Motors

Name _____ Date _____ Score _____

Objectives

Single-phase induction motors are not self-starting without some auxiliary method used to provide initial movement of the rotor. One method is the use of a starting winding. Single-phase induction motors that employ auxiliary windings for starting can have a capacitance connected in series. A capacitor-start induction motor is made to operate as though a two-phase ac source supplied the power to produce the rotating magnetic field. A capacitor is employed to provide initial starting torque and to develop a strong, uniform magnetic field.

In this activity, you will be able to analyze the operational characteristics of either a commercially manufactured capacitor motor or one that has been constructed in the laboratory. You will see the effect the capacitor has on the starting torque of the motor.

Equipment and Materials

- Multimeter

- Single-phase ac power source

- AC current meter

- Capacitor-start motor

- Tachometer

Safety

Be sure to wear eye protection while electric motors are in operation. Also, be very careful when working with high voltages.

Procedure

1. A connection diagram for a single-phase ac, capacitor-start induction motor is shown in Figure 43-1. Set up this circuit.

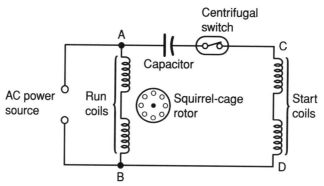

Figure 43-1. Connection diagram for a single-phase ac capacitor-start motor.

Insert Figure 43-1 here.

2. Count the number of stator coil sets (poles) that the motor you are analyzing has in the run and start circuits.

   Number of running-coil sets = _____.

   Number of starting-coil sets = _____.

3. With a meter, measure the resistance of the run windings (point A to point B on the diagram). Temporarily disconnect the run windings to make this measurement.

   R_{RUN} = _____ Ω.

4. Measure the resistance of the start windings (point C to point D on the diagram).

   R_{START} = _____ Ω.

5. Connect an ac current meter in series with one of the power lines. Apply rated voltage to the motor and observe the initial current drawn by the motor.

   Starting current = _____ A.

6. Record the running current of the motor.

   Running current = _____ A.

7. Carefully apply a load to the shaft of the motor while observing the effect on speed and current.

8. With a tachometer, measure the speed of rotation.

   Speed = _____ r/min.

9. With the motor in its original circuit configuration, perform the necessary procedure to reverse the direction of rotation. Measure and record the starting current, running current, and speed.

   Starting current = _____ A.

   Running current = _____ A.

   Speed = _____ r/min.

10. Turn off the power supply. Return all materials.

Analysis

1. Why does the addition of a capacitor in the start windings of an induction motor improve its starting torque? _____
   _____
   _____

2. Describe the following types of capacitor motors:

   a. Capacitor-start induction motor._____
      _____
      _____

   b. Capacitor-start, capacitor-run motor. _____
      _____
      _____

   c. Permanent-capacitor motor._____
      _____
      _____

3. From the data of Step 2, compare the number of starting and running coil sets of the motor. __
   _____

4. From the data of Steps 3 and 4, compare the resistances of the start and run windings of the motor. _____
   _____

5. From the data of Steps 5 and 6, compare the starting current and running current values. ____
   _____

6. How does increased load affect the speed and current of the motor? (See Step 7.) _____
   _____

7. How is the direction of rotation of the capacitor-start motor reversed? _____
   _____

10. What are some applications of capacitor motors? _____
    _____
    _____

Activity 44–Shaded-Pole AC Motors

Name _____ Date _____ Score _____

Objectives

A unique method of making a single-phase ac motor self-starting is by a process called *pole shading*. The shading poles are contained within the field poles of the machine. Generally, about one-third of the field pole is enclosed by the shading coil. When the magnetic flux around the pole starts to build up, it cuts across the shading coil. This induces a voltage into the shading coil that produces a current. The area inside of the shading coil then has a reverse polarity compared with the remaining pole face. As a result, there is a shifting effect created across the pole face from the unshaded to the shaded side. The shifting effect is significant enough to cause the rotor to turn. This turning action then makes the motor self-starting. The starting torque of the shaded-pole motor is very small, with some of the largest ratings of this type of motor around 1/20 horsepower.

The shaded-pole motor is mainly used for very small torque applications such as fan or blower motors. In this laboratory activity, you will be able to analyze either a commercially manufactured, ac shaded-pole motor or one that has been constructed in the lab.

Equipment and Materials

*   Multimeter

*   Shaded-pole ac motor

*   AC power source

*   Tachometer

*   AC current meter

Safety

Be sure to wear eye protection while electric motors are in operation. Also, be very careful when working with high voltages.

Procedure

1. The connection diagram for a shaded-pole ac motor is shown in Figure 44-1. Set up the circuit.

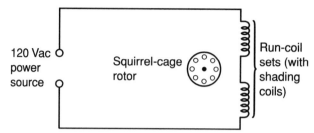

Figure 44-1. Connection diagram for a shaded-pole ac motor.

2. Measure the resistance of the stator windings.

   Winding resistance = _____ Ω.

3. Connect an ac current meter in series with the power line. Apply rated voltage to the motor and observe the initial current it draws when turned on.

   Starting current = _____ A.

4. Record the current drawn by the motor with no load applied.

   Running current = _____ A.

5. Carefully apply a load to the shaft of the motor while observing the effect on speed and current.

6. With a tachometer, measure the speed of rotation.

   Speed = _____ r/min.

7. Record the number of poles or coil sets your motor has.

   _____ poles.

8. Turn off the power supply. Return all materials.

Analysis

1. Explain how the shaded-pole induction motor produces starting rotation. _____
   _____

2. What are some applications of shaded-pole motors? _____
   _____

3. How is the synchronous speed of all ac induction motors determined? _____
   _____

4. What is the synchronous speed of the motor you analyzed? _____
   _____

5. What is meant by the term *slip* as applied to ac induction motors? _____
   _____
   _____

6. From the data of Step 6, determine the slip percentage of the motor you analyzed.

_____

_____

$$\% \text{ Slip} = \frac{\text{synchronous speed} - \text{operating speed}}{\text{synchronous speed}} \times 100 = \text{_____}$$

7. How do the starting and running currents of the shaded-pole motor compare? Why? _____

_____

_____

8. What is the effect of increased load on motor speed and current? _____

_____

_____

9. How could the direction of rotation of a shaded-pole motor be reversed? _____

_____

_____

10. What would happen if one of the shading coils became open-circuited? _____

_____

_____

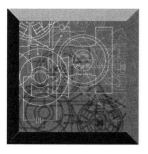

Activity 45–Three-Phase AC Synchronous Motors

Name _____ Date _____ Score _____

Objectives

Three-phase ac synchronous motors operate at a constant speed regardless of the mechanical load applied. In addition to the constant speed characteristic, this type of motor also has the ability to operate at leading or lagging power factors. By changing the dc excitation voltage applied to the rotor, the power factor at which the motor operates will vary. When a synchronous motor operates at a leading power factor, it improves the overall power factor of the system on which it is operating. This operating mode produces a capacitive effect on the system. Therefore, the machine is referred to as a *synchronous capacitor*.

In this laboratory activity, you will have an opportunity to analyze the operational characteristics of a commercially manufactured, three-phase ac synchronous motor, or one that has been constructed in the lab using a rotating machinery unit. The construction of this motor is the same as that of a three-phase ac alternator. Therefore, a three-phase alternator, such as one used in an automobile, could be substituted for an operational three-phase ac synchronous motor for a laboratory analysis.

Equipment and Materials

- Multimeter
- Three-phase ac synchronous motor
- Three-phase ac power source
- Variable dc power source
- AC current meter
- DC current meter
- Tachometer

Safety

Be sure to wear eye protection while electric motors are in operation. Also, be very careful when working with high voltages.

Procedure

1. A simplified connection diagram for a three-phase ac synchronous motor is shown in Figure 45-1. Normally, the stator coils are connected in a wye configuration. Set up this circuit.

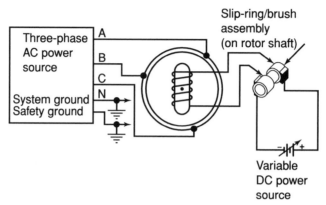

Figure 45-1. Connection diagram for a three-phase ac synchronous motor.

2. Measure the following resistance across the power lines:

    R_{A-B} = _____ Ω.

    R_{B-C} = _____ Ω.

    R_{A-C} = _____ Ω.

3. Connect the synchronous motor to a three-phase ac power source in order to supply the rated voltage to the stator coils of the motor.

4. Place an ac current meter in series with one of the power lines.

5. Connect the terminals from the slip-ring/brush assembly of the rotor to a variable dc power source. Adjust this dc voltage to zero.

6. Determine the number of stator-coil sets (poles) that your motor has.

    Stator coil sets = _____.

7. Apply three-phase ac power to the machine. Does it start? _____.

8. If the machine does not start, and if it does not have damper windings, it may be necessary to bring its speed up to near its synchronous speed by connecting another motor to its shaft to supply the necessary speed.

9. With the synchronous motor operational, record the ac current indicated on the meter.

    Current = _____ amperes ac.

10. Gradually increase the dc voltage applied to the rotor while observing the effect on speed and ac current.

11. Turn the dc voltage down to zero once more. Connect a dc current meter in series with the power source.

12. Now measure the sped of rotation with a tachometer.

    Speed = _____ r/min.

13. If available, record the rated dc voltage of the machine you are analyzing.

    _____ volts dc at _____ power factor.

14. Vary the dc voltage applied to the rotor in increments of approximately 2 volts until about 50% above the rated dc voltage is reached. At each dc voltage value, record the dc current through the rotor and the ac current drawn by the stator. Record your data in Figure 45-2.

DC voltage	Rotor current (DC)	Stator current (AC)
0		

Figure 45-2. Rotor current versus stator current.

15. With the rated voltage applied to the rotor, measure the speed of rotation with a tachometer.

    Speed = _____ r/min.

16. Carefully apply a load to the shaft of the machine. Observe the effect of increased load on ac stator current, dc rotor current, and speed.

17. Turn the dc voltage back to zero volts dc. Now gradually increase the dc voltage until the rotor is synchronized in step with the revolving stator field.

18. Turn off the dc, then the ac power source. Return all materials.

Analysis

1. Describe the basic construction of a three-phase synchronous motor. _____
   _____

2. What is the speed of the following three-phase synchronous motors?

    6 pole* = _____ r/min.

    12 pole* = _____ r/min.

    24 pole* = _____ r/min.

    *Refers to the total number of stator poles.

3. How does a three-phase synchronous motor differ from a three-phase induction motor? ____
   _____

4. What are some methods of starting three-phase synchronous motors? _____
   _____

5. What is pull-in torque? Pull-out torque? _____
   _____

6. When is a three-phase synchronous motor considered to be underexcited? Overexcited? ____
   _____

7. Verify your observation of the number of stator poles in Step 6 with the speed of rotation measured in Step 15.

_____

_____

$$\text{Synchronous speed (r/min.)} = \frac{\text{ac frequency} \times 120}{\text{number of poles/phase}} = \underline{\qquad}$$

8. What effect does increased rotor current have on the stator current of the synchronous motor?

_____

_____

9. What effect does increased load have on: (See Step 16.)

a. speed? _____

_____

b. dc rotor current? _____

_____

c. ac stator current? _____

_____

10. What effect does increasing the dc rotor voltage (from zero) have on the speed of the motor?

_____

11. With the data recorded in Figure 45-2, make a graph of dc rotor current versus ac stator current. Use Figure 45-3. Also mark the point on the curve that corresponds to the highest power factor operation of the motor.

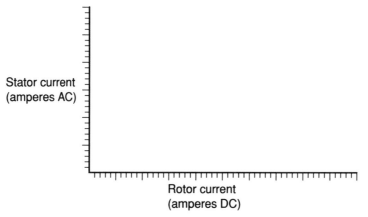

Figure 45-3. Graph of dc rotor current versus ac stator current.

Activity 45—Three-Phase AC Synchronous Motors

Activity 46–AC Synchronous Motors

Name _____ Date _____ Score _____

Objectives

The ac synchronous motor used in automated industrial systems is commonly classified as a constant-speed device. It has extremely rapid starting, stopping, and reversing characteristics. Motors of this type only necessitate a simple clockwise/stop/counterclockwise rotary switch for control. The starting and running currents of a synchronous motor are identical, which is unique for ac motors. This characteristic means that a motor of this type can withstand a high inrush of current when direction changes occur. As a general rule, the synchronous motor can even be stalled without damaging the motor.

Motors of this type are commonly used as drive mechanisms for industrial systems in machinery operations. Starting begins within one and one half cycles of the line frequency, and stopping occurs in five mechanical degrees of rotation. This motor represents a unique part of all servomechanisms used in precision rotary control applications today.

In this activity, a simple test circuit will be constructed so that you can observe an ac synchronous motor's starting, stopping, and current flow characteristics.

Equipment and Materials

- AC ammeter—0–5 A
- Ac synchronous motor (Superior Electric Type SS150 or equivalent motor rated at no more than 200 in/oz.)
- Capacitor—3.75 µF, 330 Vac
- Resistor—250 Ω, 25 W
- SPST switches (3)
- Isolated ac power source—120 V
- Piece of wood

Safety

Be sure to wear eye protection while electric motors are in operation. Also, be very careful when working with high voltages.

Procedure

1. Construct the ac synchronous motor test circuit of Figure 46-1. Ensure that the motor is mounted securely.

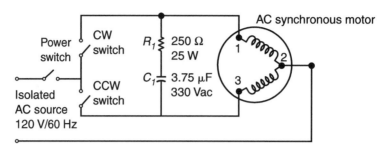

Figure 46-1. AC synchronous motor test circuit.

2. Turn on the CW switch for clockwise rotation.

3. Momentarily turn on the power switch and observe the rotation of the motor.

4. With a tachometer measure the speed of rotation.

   Speed = _____ r/min.

5. Momentarily turn off the power switch and observe the stopping action of the synchronous motor. You may want to run several trials to see the quickness of the stopping action. How does this compare with other ac motors? _____

   _____

6. Note each time that the motor is turned on, how quickly it comes to speed.

7. With an ac ammeter, measure the running current and starting current of the synchronous motor.

   Running current = _____ A.

   Starting current = _____ A.

8. Carefully wedge a piece of wood between the rotating shaft and the bench while holding the motor. This will provide a simple loading method for test purposes.

9. When the motor is loaded down, how does the running current respond? _____

   _____

10. If the motor is completely stalled, how does the running current respond? _____

    _____

11. Remove the load from the motor, turn off the clockwise switch, and turn on the counterclockwise switch. The motor should rotate equally well in the counterclockwise direction and have the same basic characteristics. Test these again to verify the theory.

12. Switch off the counterclockwise switch, then switch on the clockwise switch and notice the ease with which direction change occurs.

13. Turn off the ac power source and disconnect the circuit. Return all materials.

Activity 46—AC Synchronous Motors

Analysis

1. Discuss the operation of an ac synchronous motor. _____

_____

_____

_____

_____

_____

_____

_____

_____

_____

_____

_____

_____

_____

_____

_____

Activity 47–Synchro Systems

Name _____ Date _____ Score _____

Objectives

Synchro systems are commonly classified as two or more generator/motor units connected in such a way that they permit the transmission of angular shaft changes by electromagnetic field changes. When the shaft of the generator unit is turned, it automatically causes a corresponding change in the motor shaft position. Through systems of this type, it is possible to achieve accurate control over a distance.

Typical synchro systems contain two or more electromagnetic devices that are similar in appearance to a small electric motor. These devices are, however, connected in such a way that the angular position of the generator shaft is transmitted through interconnecting wires to the motor or receiving unit.

When electrical energy is first applied to a generator or motor unit, the unit will move to a specific location, then come to rest. This position is called electrical 0°. Both synchro generators and motors are adjusted to 0° in the same manner.

In this activity, you will connect the motor and generator units of a simple synchro system together to test their operation. Through this experiment, you will become familiar with actual circuit terminal designations and be able to connect the units together to achieve clockwise and counterclockwise rotation.

Equipment and Materials

- Multimeter
- Oscilloscope
- Isolated variable ac power source—0–120 V
- Variable ac transformer
- Synchro generator/motor set

Safety

Be sure to wear eye protection while electric motors are in operation. Also, be very careful when working with high voltages.

Procedure

Section A: Electrical 0° Determination

1. Position the generator part of the synchro system so that you can readily observe the terminal connections at the rear of the unit. The stator terminals are usually labeled S_1, S_2, and S_3. The rotor terminals are generally labeled R_1 and R_2.

2. Connect the isolated variable ac power source to the rotor terminals as indicated in Figure 47-1.

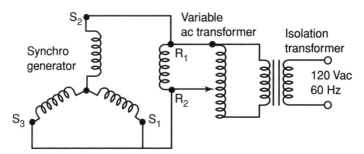

Figure 47-1. Circuit for determining the electrical 0° position of a synchro generator unit.

3. Adjust the variable ac transformer to its lowest output voltage level. Then turn on the ac source and gradually increase the voltage until the rotor winding locks in place. This represents the electrical 0° position of the rotor.

4. Prepare the multimeter to measure ac voltage and connect it across R_1 and R_2.

   The supply voltage measures _____ volts ac.

5. With the rotor locked at 0°, adjust the indicating dial to the mechanical 0° position.

6. Another way to determine the electrical 0° position of a synchro generator is using the voltmeter method. Turn off the power source and connect the generator unit so that S_1, S_2, and S_3 are open with respect to the ac source applied to the rotor. The stator windings in this case are energized entirely by induction from the rotor.

7. Turn on the power source and measure the voltage across S_1 and S_2. Turn the rotor while observing the voltage. Note that zero voltage occurs at two locations during one revolution. This represents electrical 0° and 180°.

8. To distinguish between 0° and 180°, turn off the power source and connect a jumper wire between R_1 and S_2. Then turn on the power source and measure the voltage across R_2 and S_1. Turn the shaft or dial and note that the voltage has a high value at one position and a lower value at an opposite position. Electrical 0° and the low point are the same. The 180° position is represented by the highest voltage value.

9. Turn off the power source and disconnect the generator unit.

10. Connect the motor and receiver unit in the same manner outlined for the generator, and set its electrical 0° indicator.

11. Turn off the power source and disconnect the circuit.

Section B: Synchro Position Control

1. Connect the synchro generator/motor units together as indicated in Figure 47-2.

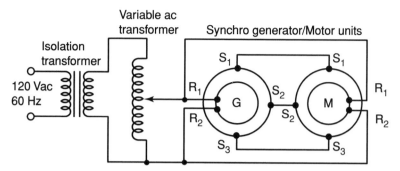

Figure 47-2. Circuit for observing position control of a synchro generator/motor system.

2. Turn on the power source and increase the voltage of the variable ac transformer to the rated rotor voltage. As a general rule this is usually 120 volts ac.

3. Move the generator rotor to the 0° position. The motor unit should follow or track the generator.

4. Rotate the generator 90° in the clockwise direction.

   How does the motor respond? _____

   Return the generator to the 0° position and then rotate the generator 90° counterclockwise.

   How does the motor respond? _____

5. You may want to try a variety of position changes to confirm your answer.

6. Prepare the oscilloscope for operation and connect the ground lead to R_2 and the probe lead to R_1. Turn on the power source and turn on the rotor through one complete revolution in each direction. How does the waveform respond to this physical change in the rotor? _____

7. Turn off the power source and connect the ground lead of the oscilloscope to S_1 and the probe lead to S_3. Turn on the ac power source and turn the rotor one revolution in each direction while observing the waveform. How does it respond? _____

8. Turn off the power source and reverse the R_1 and R_2 rotor leads of the motor unit. Turn on the power source and test the operation. How has the circuit changed, compared with previous steps? _____

   _____

9. Turn off the power source and return R_1 and R_2 as they were originally connected in Figure 47-2. Connect stator leads S_1 of the generator to S_3 of the motor and S_3 of the generator to S_1 of the motor.

10. Turn on the power source and test the rotary control action of the generator/motor set. How does it respond when connected in this manner? _____

11. Turn off the power source and disconnect the circuit. Return all materials.

Analysis

1. Explain how the stator windings are energized by inductance as noted in Section A, Step 6.

   _____

2. What is meant by the term *electrical 0°*? _____

   _____

3. What is the electrical difference between a synchro generator and a synchro motor of the same
   set? _____

   _____

4. What does the waveform of Section B, Step 6 show about the stator voltage? _____

   _____

5. Why does the motor rotor track the generator rotor of a synchro system? _____

   _____

Activity 48–Industrial Symbols and Diagrams

Name _____ Date _____ Score _____

Objectives

In Activity 4, basic electrical symbols were reviewed. There are several symbols that are unique to industrial applications. In this activity, you will examine the common industrial symbols and diagrams shown in Figure 48-1.

Resistor	R
Variable resistor	R (with arrow)
Lighting panel	▬
Power panel	P
Transformer	T
Junction box; 4-inch standard octagon box	J
Conduit (joined)	—‖—
Incandescent light	⊗
Mercury-vapor light	⊗ Hg
Fluorescent light	▭
Single-pole switch	S₁
Two-pole switch	S₂
120-volt convenience receptacle	—⊖—

Special receptacle	◁
Conduit (concealed)	– – – –
Conduit (exposed)	———
Ground wire	—G—
Air circuit breaker	⌒
Current transformer	⌒⌒
Fuse	⊏□⊐
High-voltage primary fuse cutout	—
Lightning arrester	—o o—
Two-winding transformer	⦚⦚
Autotransformer	⦚
Solenoid	—⋀—
Push button (momentary open)	o ⊥ o

Push button (momentary close)	
Push button (momentary open, closed)	
Pressure-actuated switch	
Temperature-actuated switch	
Thermal cutout	—⌒⌒—
Open switch with time delay on closing	
Closed switch with time delay on opening	
Ground connection	⏚
Normally closed contact	⊣/⊢
Normally open contact	⊣ ⊢
Time delay on closing	⊣ ⊢
Time delay on opening	⊣/⊢

Figure 48-1. Industrial schematic symbols.

193

Industrial Diagrams

The most common diagrams used to illustrate the components, circuits, and subsystems of an industrial system are connection diagrams, block diagrams, schematic diagrams, and one-line diagrams.

In this activity, you will need to do some research to locate some simple industrial systems that can be represented by using the appropriate type of diagram.

Analysis

In the space next to each listed device, draw the proper symbols for that device.

1. Incandescent lamp:

2. Mercury vapor lamp:

3. Fluorescent lamp:

4. Circuit breaker:

5. Fuse:

6. Push-button switch (normally open):

7. Push-button switch (normally closed):

8. Pressure-actuated switch:

9. Temperature-actuated switch:

10. Thermal cutout:

11. Normally closed contact:

12. Normally open contact:

Activity 48—Industrial Symbols and Diagrams

13. *Connection Diagram*: The connection diagram shows the physical relationships of the parts of a system. It is used to identify various parts for installation or servicing. The wires that interconnect circuits are ordinarily shown in their correct location. In the following space, draw a simple connection diagram for an industrial system.

14. *Block Diagram*: The block diagram is a very general method used to show how subsystems of any industrial system fit together to form a functional unit. Rectangles are used to represent subsystems and arrows are used to connect the blocks and show the interrelationships of subsystems. In the following space, draw a block diagram of an industrial system.

15. *Schematic Diagram*: The schematic diagram is used to show the electrical circuit relationships of the various components of an operational system. The electrical, but not necessarily the physical, layout of the circuit is shown. This type of industrial diagram is the most often used. In the following space, draw a schematic diagram of an industrial circuit.

16. *One-Line Diagram*: The one-line diagram is a simplified form of the schematic diagram. This type of diagram is used to show how the components of a system fit together. It is similar to the block diagram in several respects. However, actual symbols, rather than rectangular shapes, are used to show component parts. In the following space, draw a one-line diagram of an industrial system.

Activity 48—Industrial Symbols and Diagrams

Activity 49–Types of Switches

Name _____ Date _____ Score _____

Objectives

In this activity, you will learn to correctly identify the types of switches that are used in electrical control systems. It is important for a technician to be able to identify switch types since they are the most basic type of electrical control.

Procedure

1. Study the symbols of switches in Chapter 13 of *Electrical Motor Control Systems*.

2. Learn the correct names of all types of switches.

3. Learn the uses of the switches as you study them.

4. The symbols of several common types of switches are shown in Figure 49-1. Fill in the correct names of the switches.

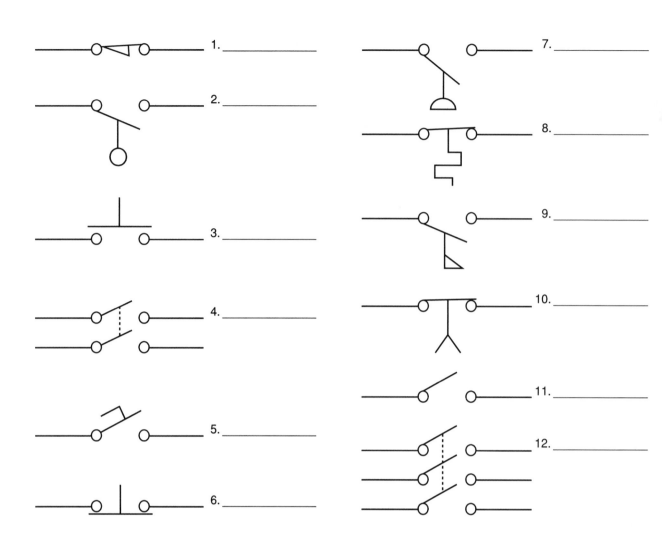

1. _____

2. _____

3. _____

4. _____

5. _____

6. _____

7. _____

8. _____

9. _____

10. _____

11. _____

12. _____

5. Your instructor can now give you a test to see how many of the types of switches available in your laboratory or shop you can identify.

Analysis

1. For each of the switches shown in Figure 49-1, write a brief description of the use of that type of switch.

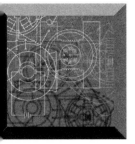

Activity 50–Float Switches

Name _____ Date _____ Score _____

Objectives

A float switch is an electromechanical control switch connected to a rod or chain that moves in response to the rise and fall of liquid in a tank or reservoir. Depending on its application, the normally open or normally closed contacts of the switch can be used.

Float switches are used on sump pumps, fill tanks, and flow lines. With the use of float switches, the automatic control of a pump motor, motor driven gate valve, or other similar device can easily be arranged.

In this activity, a lever-type limit switch can be used in place of the regular float switch. The operation of the limit switch and the float switch are identical. The lever arm of the limit switch is usually activated by a moving object while the float ball and lever arm of the float switch are activated by the level of liquid within the container.

The use of normally open or normally closed contacts is determined by whether the motor is to operate on the filling or emptying of a reservoir. In this activity, you will examine the operation of a float switch in both full-tank and sump operation.

Equipment and Materials

Lever-type limit switch or float switch

Small horsepower three-phase motor*

Single coil three-phase magnetic contactor*

Three-phase power supply*

*This activity can also be done with single-phase ac power.

Safety

Be very careful when working with high voltages.

Procedure

1. Connect the circuit of Figure 50-1. As connected, this is a full-tank operation.

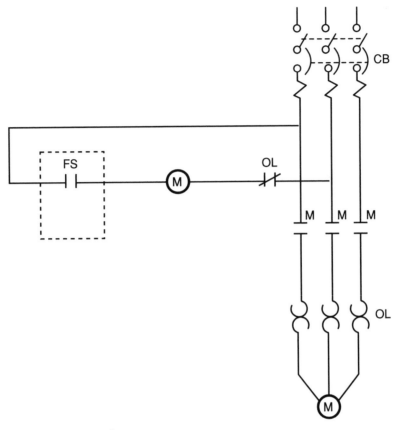

Figure 50-1. Float switch motor control.

2. Have the instructor check your circuit before applying power.

3. Being a full-tank operation, the normally open contacts are used. Pull the lever of the switch to one side. As long as the lever is held, the motor should run.

4. Release the lever. What happens? _____

_____

5. What condition of liquid level does the depressed lever represent? _____ What condition does the released lever represent? _____

6. Disconnect the power. Alter the circuit of Figure 50-1 so that the circuit operates as a sump pump operation. What must be changed to accomplish this? _____

7. Reconnect the power. Does the motor run now? _____

8. Pull the lever to one side. What happens? _____

9. Release the lever. What happens? _____

_____

10. Disconnect the circuit and turn off the power. Return all materials.

Analysis

1. Write a brief explanation of the operation of both circuits in this activity. Also, explain what is meant by *full-tank* and *sump operation*. Include your own observations and conclusions about this activity in your discussion. _____

_____

_____

_____

_____

_____

_____

_____

_____

Activity 51–Motor Control with Magnetic Contactors

Name _____ Date _____ Score _____

Objectives

A very common method used to turn electric motors on or off is a magnetic contactor used in conjunction with a set of push-button switches. A magnetic contactor is a type of relay. It is energized by applying a voltage to its electromagnetic coil. This control voltage causes a solenoid action that closes the contacts of the device. These contacts can be connected in series with the power line going to a motor to cause it to turn on. The coil of the contactor is a relatively high resistance, causing it to draw a small current from the power line. This small current is capable of energizing the relay so that the closing of its contacts will apply power to a motor. Thus, a small current is used to control the larger current of the motor. The surface area of the contacts is usually fairly large to accommodate the current level drawn by the motor.

In this activity, you will use a magnetic contactor and push-button switch assembly to control the operation of a motor. Be sure to note the difference in magnitude of the control current through the contactor coil and the current drawn by the motor.

Equipment and Materials

- Multimeter
- Single-phase ac power source
- Single-phase ac motor
- Start/stop push-button station
- Connecting wires

Safety

Be sure to wear eye protection while electric motors are in operation. Also, be very careful when working with high voltages.

Procedure

1. Construct the circuit shown in Figure 51-1.

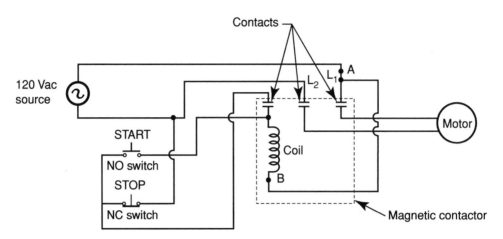

Figure 51-1. Magnetic contactor motor control circuit.

2. Press the normally open push-button switch (START switch) to test the circuit for operation. The relay coil should energize, causing the motor to rotate.

3. Press the normally closed push-button switch (STOP switch). The motor should stop running.

4. Turn off the 120 volts ac power source and connect an ac ammeter in series with the ac line at point A.

5. Apply power to the circuit, press the START switch, and measure the starting and running current of the motor.

   Starting current = _____ amperes ac.

   Running current = _____ amperes ac.

6. Turn off the circuit and disconnect the ac power source. Reconnect the ac ammeter in series with the relay coil at point B.

7. Apply power to the circuit, press the START switch, and measure the current through the coil.

   Coil current = _____ amperes ac.

8. With the ac line plugged into a variable 0–120 volt power source, perform the following steps (if variable ac is available).

9. Set the power source to zero volts and then turn it on. Keep the START button depressed while increasing the applied voltage. Measure and record the "pull-in" current and voltage of the contactor.

   Pull-in current = _____ amperes ac.

   Pull-in voltage = _____ volts ac.

10. Gradually reduce the applied voltage until the contacts open. Record the "drop-out" current and voltage.

   Drop-out current = _____ amperes ac.

   Drop-out voltage = _____ volts ac.

Analysis

1. Why are magnetic contactors used to control motors? _____

_____

2. Explain the operation of a magnetic contactor circuit to control a single-phase motor. _____

_____

3. Draw a schematic diagram showing how a magnetic contactor could be used to control a three-phase motor.

4. What is meant by "pull-in" current? _____

_____

5. What is meant by "drop-out" current? _____

_____

Activity 52–Magnetic Contactor Wiring

Name _____ Date _____ Score _____

Objectives

During this activity, you will demonstrate your ability to correctly wire a magnetic contactor control.

Equipment and Materials

- Multimeter
- Single-phase ac power source
- Single-phase ac motor
- Start/stop push-button station
- Connecting wires

Safety

Be sure to wear eye protection while electric motors are in operation. Also, be very careful when working with high voltages.

Procedure

1. A control panel for a magnetic contactor used to control an electric motor is shown in Figure 52-1. This is a pictorial diagram of the laboratory activity circuit.

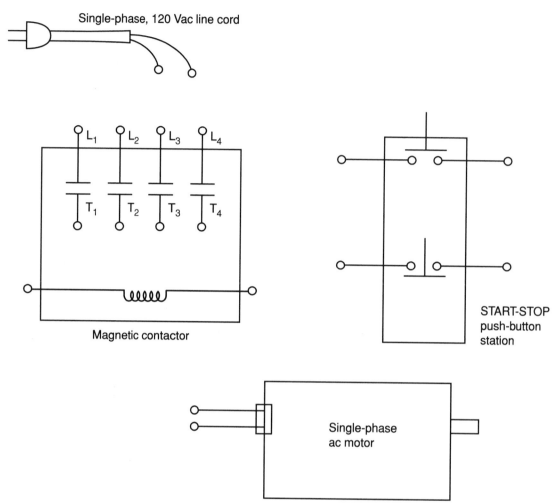

Figure 52-1. Control panel for a magnetic contactor.

2. Make a sketch of the *control diagram* for starting and stopping a single phase, 120 Vac motor below.

Activity 52—Magnetic Contactor Wiring

3. Complete the *wiring diagram* of the pictorial below to correspond to your control diagram. Use Figure 52-1.

4. Wire a magnetic contactor, start push-button, and stop push-button to control a single-phase ac motor.

5. Have your completed circuit checked for proper operation.

   Instructor's Approval: _____

Activity 53–Electromagnetic Relays

Name _____ Date _____ Score _____

Objectives

Relays are electromagnetic switches and are excellent examples of how a magnetic field attracts a magnetic material. These devices contain a coil that creates an electromagnetic field, an armature that is constructed of a magnetic material attracted by the coil, and a number of contacts or switches that open or close when the magnetic field attracts the armature.

In this activity, you will study the electromagnetic characteristics of a relay.

Equipment and Materials

Multimeter
- Multicontact relay
- Lamp with socket—6 volt
- Variable dc power supply
- Resistor—1 kΩ
- Battery—6 volt
- Connecting wires

Procedure

1. Prepare the multimeter to measure resistance. Measure and record the resistance of your relay coil. _____ Ω.

2. Using the multimeter, determine how many normally open and normally closed contacts are used with your relay.

   Number of normally open contacts = _____.

   Number of normally closed contacts = _____.

3. Construct the circuit illustrated in Figure 53-1. Be sure that the variable dc power supply is adjusted to zero. The multimeter should be adjusted to measure dc current on the highest range.

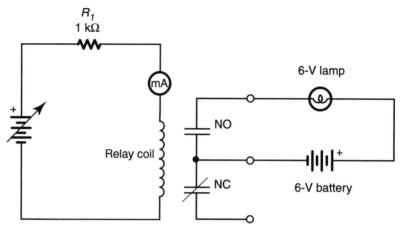

Figure 53-1. Circuit using electromagnetic relay.

4. Slowly adjust the variable dc power supply from zero until the 6-volt lamp is turned on. Record the current measured by the multimeter when the relay is energized. This is the pickup current.

   Pickup current = _____ mA.

5. Slowly adjust the variable dc power supply toward zero until the 6-volt lamp is turned off. Record the current measured by the multimeter when the relay de-energized. This is the dropout current.

   Dropout current = _____ mA

6. Turn the variable power supply off.

7. Alter the circuit to match the circuit in Figure 53-2.

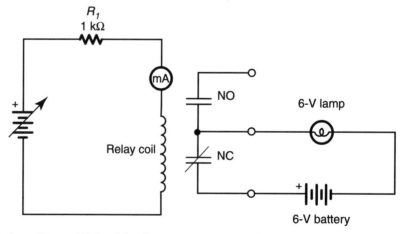

Figure 53-2. Circuit from Figure 53-1 with relay contact reversed.

Activity 53—Electromagnetic Relays

8. You will notice that the only difference in the two circuits is the type of contacts used. In Step 3, the normally open contacts were used. In this procedure the normally closed contacts are used, causing the lamp to remain on until the relay is energized.

9. Adjust the variable dc power supply and record the pickup and dropout currents as you did in Steps 4 and 5.

   Pickup current = _____ mA

   Dropout current = _____ mA

10. How do the currents recorded in Step 9 compare with the current recorded in Steps 4 and 5?

   _____

11. How did the action of the 6-volt lamp in Steps 4 and 5 compare with the action of the lamp in Step 9?

   _____

Analysis

1. What are normally open contacts? _____

   _____

2. What are normally closed contacts? _____

   _____

3. What is meant by the term pickup current? _____

   _____

4. What is meant by the term dropout current? _____

   _____

5. Using Ohm's law, compute the voltage across the relay coil when the relay is energized. (See Steps 1 and 4.)

   $V = I \times R =$ _____ V

Activity 54–Load Centers

Name _____ Date _____ Score _____

Objectives

Electrical power used to operate electrical devices is often generated from coal-fired steam systems, hydroelectric systems, or nuclear-powered systems. The power is carried across the country by high voltage transmission lines. The voltage is then reduced to a usable value by means of a step-down transformer. The power enters a building through a service entrance that is connected to a metering device, through a main disconnect, and into the load center.

The load center is commonly known as a distribution panel, since this is where power is distributed to individual circuits. The load center is protected by the main circuit breaker, which breaks the hot power lines when current values are above the breaker rating. The neutral power line is not broken since it is connected to a grounding bar that is connected to earth ground. From the main breaker the power is fed to bus bars from which the individual circuits are taken. Each circuit is protected by a circuit breaker with a current rating designated by the circuit.

In this activity, you will set up and examine the circuit for a distribution panel, or load center.

Equipment and Materials

Single-phase safety switch

Single-phase load centers (2)

Circuit breakers—15 amp, 20 amp, 30 amp, 120/240 volt

Electrical conductors

Wireway

Safety

Be very careful when working with high voltages.

Procedure

1. Properly wire the control panel shown in Figure 54-1.

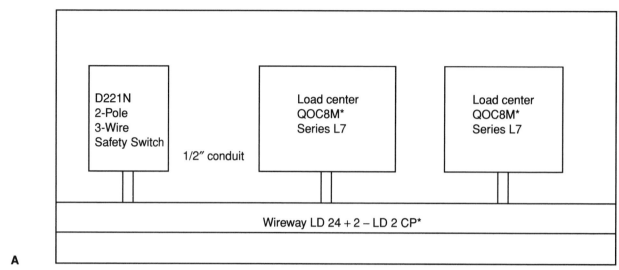

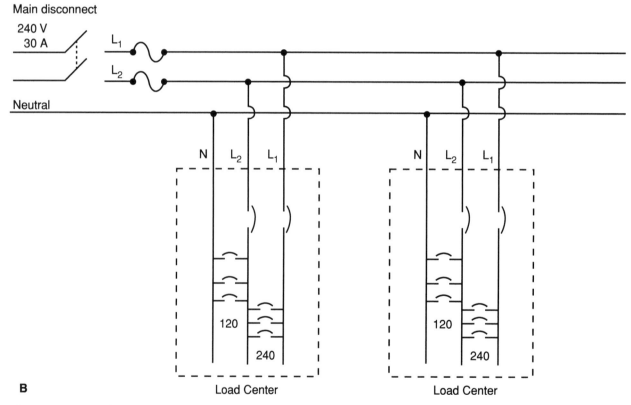

Figure 54-1. Load center controls. A—Layout of suggested laboratory control panel. B—Schematic of circuit wiring of control panel.

2. Have your completed circuit checked for proper operation.

Instructor's Approval: _____

Analysis

Discuss the following types of control equipment.

1. Load center.

2. Circuit breakers.

3. Single-phase safety switch.

4. Wireways (raceways).

Activity 55–Three-Phase Motor Controls

Name _____ Date _____ Score _____

Objectives

The panel of Figure 55-1 consists of a manual contactor, a magnetic contactor, and a combination starter. A manual starter is operated by human operators in most cases. The switch is spring loaded to prevent excessive arcing of the contacts. This is the simplest contactor.

The magnetic contactor is operated by means of electrical energy, which is achieved by means of a start/stop control station. When the start button is pressed, current can flow from L2 through the close stop contacts, through the momentarily closed start contacts, throughout the coil and to L2. This completes the coil circuit and closes the line and holding contacts. When the start button is released, the coil stays energized through the holing contacts, which are in parallel with the start button contacts. A pilot light is connected in parallel with the coil, so anytime the coil is energized, the pilot light is energized. Since the stop button contacts are in series with the coil, the circuit can be opened by pressing the stop button.

The combination starter has the overload protection (which is a special fuse) and circuit breaker. A current-limiting fuse interrupts the current before the short-circuit can occur. On low-value short-circuit currents, the breaker trips before the fuse opens. If the short-circuit current exceeds the capabilities of the circuit breaker, the fuse opens the circuit and protects the breaker from damage. All lines are opened when either devices trip. This type of protection is valuable where extremely high currents flow under short-circuit conditions.

The combination starter has a start-stop station, overload devices, and magnetic contactors enclosed in a single enclosure. When the start button is pressed, the coil circuit is completed which energizes the coil and closes the line and holding contacts. The stop button is in series with the coil and when pressed, it opens the coil circuit.

In this activity, you will set up and examine a three-phase motor control panel.

Equipment and Materials

- Three-phase safety switch
- Three-phase manual starter
- Three-phase magnetic starter
- Combination starter
- Wireway
- Start/stop push-button station
- Melting-alloy thermal overload (6)

Safety

Be very careful when working with high voltages.

Procedure

1. Properly wire the control panel shown in Figure 55-1, using the equipment provided to study three-phase motor controllers.

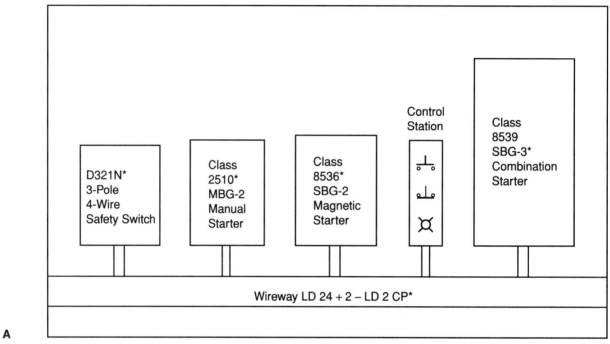

*These are sample manufacturers' numbers

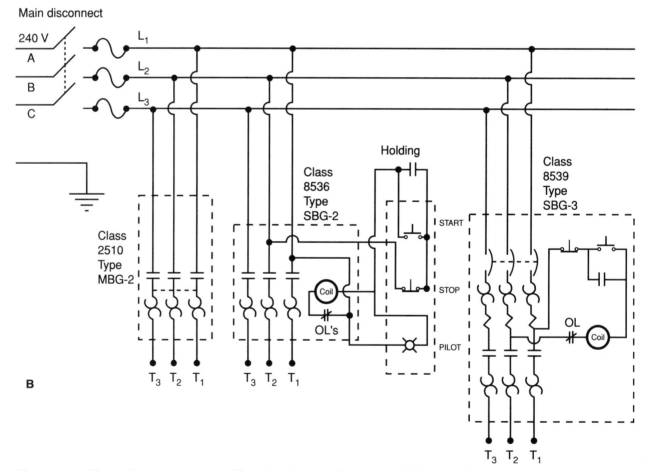

Figure 55-1. Three-phase motor controllers. A—Layout of suggested lab control panel. B—Schematic.

Activity 55—Three-Phase Motor Controls

2. Have your completed circuit checked for proper operation.

Instructor's Approval: _____

Analysis

Discuss the following types of control equipment.

1. Three-phase safety switch.

2. Three-phase manual starter.

3. Three-phase magnetic starter.

4. Combination starter.

5. Overload devices.

    a.   Melting alloy.

    b.   Bimetallic.

Activity 56—Three-Phase Reversing Controls

Name _____ Date _____ Score _____

Objectives

The panel of Figure 56-1 demonstrates the reversing controls for three-phase machines. These systems are used to reverse the direction of motor rotation by electrically interchanging the position of any two of the three conductors supplying the motor. A three-phase motor is the simplest of any motor to reverse. A direct short would result between phases if both the forward and reverse contactors of the starter were in the energized position at the same time. To prevent this, the contactors are mechanically interlocked so only one or the other can be closed. Three-phase motors can also be reversed by drum controllers. The drum controller is manually operated and the mechanical interlocking is inherent in the construction.

The operation of this circuit is controlled by the forward/reverse/stop push-button station that can be located at any convenient location. The control circuits are operated from 120 volts, which is supplied from a step-down transformer. When the forward button is pressed, contacts are momentarily closed, allowing current to flow through the forward coil, which closes the magnetic contactors and the holding or auxiliary contacts that keep the circuit closed after the forward button is released. When the reverse button is pressed, the forward coil circuit is broken and the reverse coil circuit is closed, allowing current to flow through the reverse coil, which closes the reverse contactors and its auxiliary contacts. Either coil circuit can be opened by the stop button, which opens the circuit to the secondary side of the transformer. A maintain/stop button, which can be placed at any convenient place, is in series with the stop of the forward/reverse station. This button is a manual push-pull that has a lamp circuit that is closed when in the pressed position. Another way of opening the forward and reverse circuit is by means of limit switches. A limit switch is a device that is normally closed and opened by mechanical means.

In this activity, you will set up and examine the circuit for three-phase reversing controls.

Equipment and Materials

- Three-phase safety switch
- Limit switch (2)
- Emergency stop push-button switch
- Three–phase ac power source
- Three-phase reversing control
- Start/stop/reverse control station
- Transformer—2:1 ratio
- Drum controller
- Wireway
- Conduit—1/2 inch EMT
- Conductors—#12 AWG

Safety

Be very careful when working with high voltages.

Procedure

1. Properly wire the control panel shown in Figure 56-1.

2. Have your completed circuit checked for proper operation.

   Instructor's Approval: _____

Analysis

Discuss the following types of control equipment:

1. Three-phase forward/reverse contactor

2. Control transformer

3. Push-button head designs

4. Drum controller

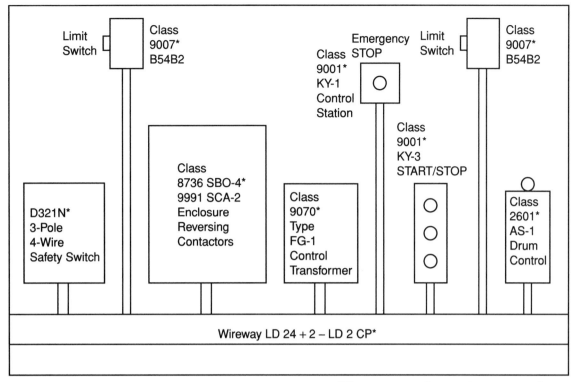

Limit
Switch

Class
9007*
B54B2

D321N*
3-Pole
4-Wire
Safety Switch

Class
8736 SBO-4*
9991 SCA-2
Enclosure
Reversing
Contactors

Class
9070*
Type
FG-1
Control
Transformer

Emergency
STOP
Class
9001*
KY-1
Control
Station

Class
9001*
KY-3
START/STOP

Limit
Switch

Class
9007*
B54B2

Class
2601*
AS-1
Drum
Control

Wireway LD 24 + 2 – LD 2 CP*

A

*These are sample manufacturers' numbers

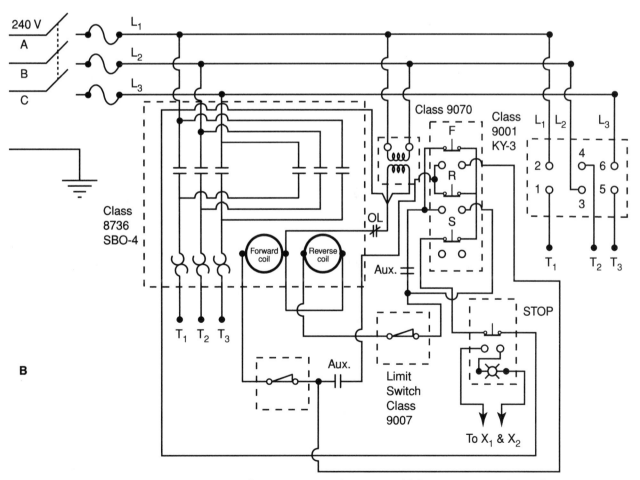

Main disconnect

240 V

A
B
C

L_1
L_2
L_3

Class
8736
SBO-4

Forward
coil

Reverse
coil

OL

Aux.

T_1 T_2 T_3

Class 9070

Class
9001
KY-3

F
R
S

L_1 L_2 L_3

2   4   6
1   3   5

T_1 T_2 T_3

STOP

Aux.

Limit
Switch
Class
9007

To X_1 & X_2

B

Figure 56-1. Three-phase reversing controls. A—Layout of suggested laboratory control panel.
B—Schematic.

Activity 56—Three-Phase Reversing Controls

225

Activity 57–Single-Phase Control Equipment

Name _____ Date _____ Score _____

Objectives

The panel of Figure 57-1 demonstrates some of the many possibilities of single-phase control systems. On this panel, a motor can be delayed from the time the selector switch is activated until some time later. This type of motor control is used in sequencing control. When the selector switch is activated, a magnetic coil is energized, pulling the iron core in and allowing the dashpot type timer to time out. This closes the timer contacts allowing the coil to the control relay to be energized, closing the control relay contacts that allow power to the motor.

Another circuit is the temperature control relays. When the temperature reaches a predetermined temperature, the temperature control contacts close by means of expanding liquid, causing the coil of a control relay to energize. This closes the control relay contacts, which allows power to flow to the motor.

A control relay can also be used as a magnetic contactor. This is done by using two push buttons in conjunction with a control relay. The coil is in series with normally open start contacts. When these contacts are closed, the coil circuit is completed through the normally closed stop contacts. At this time, the contacts are closed. Now there is a set of holding or maintaining contacts that are mechanically connected to the contactor and close at the same time. This contact parallels the normally open start contact and allows the coil to stay energized after the normally open contacts are released. The circuit can be turned off by depressing the stop button, which opens the circuit to the coil.

In this activity, you will set up and examine a single-phase control circuit.

Equipment and Materials

Single-phase ac power source

Single-phase safety switch

Time delay relay

Temperature switch

Pressure switch

Control relay (2)

Start/stop control station

Conduit—1/2 inch EMT

Conductors—#12 AWG

Wireway

Safety

Be very careful when working with high voltages.

Procedure

1. Properly wire the control panel shown in Figure 57-1, using the equipment provided to study single-phase control systems.

2. Have your circuit checked for proper operation.

   Instructor's Approval: _____

Analysis

Discuss the following types of control equipment.

1. Time-delay relay.

2. Temperature switch.

3. Pressure switch.

4. Control relay.

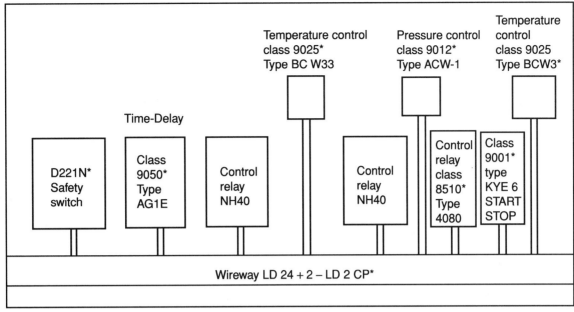

A

*These are sample manufacturers' numbers

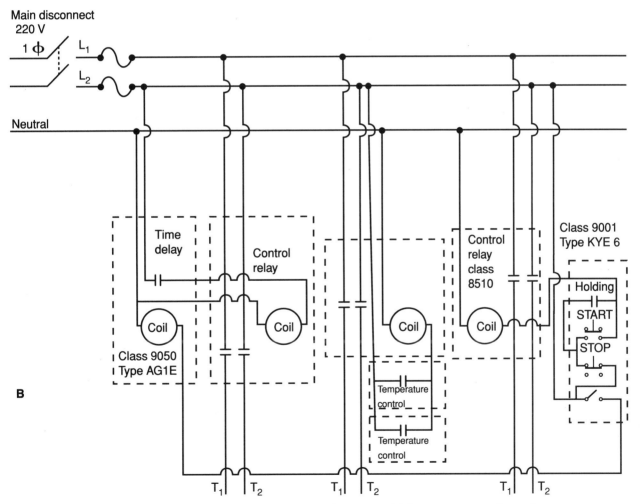

B

Figure 57-1. Single-phase control systems. A—Layout of suggested laboratory control panel. B—Schematic.

Activity 58–Forward and Reverse Control Circuits

Name _____ Date _____ Score _____

Objectives

Most types of electrical motors can be made to rotate in either direction by simple modifications of their connections. Ordinarily, motors used in industry require two magnetic contactors to accomplish forward and reverse operation. These contactors are used in conjunction with a set of three push-button switches—FORWARD, REVERSE, and STOP. When the FORWARD push-button switch is depressed, the forward contactor will be energized. It is deactivated when the STOP push-button switch is depressed. A similar procedure takes place during reverse operation.

In this activity, you will observe the forward and reverse operation of: (1) a dc shunt motor, and (2) a three-phase induction motor. The activity is designed for observation only. Therefore, no measurements will be taken. Keep in mind the function of the magnetic contactor as a low-current control device for controlling the large currents drawn by motors.

Equipment and Materials

DC shunt motor

Three-phase induction motor

Push-button switches, normally open (2)

Push-button switch, normally closed

Magnetic contactors (2)

DC power source

Three-phase ac power source

Safety

Be sure to wear eye protection while electric motors are in operation. Also, be very careful when working with high voltages.

Procedure

1. Construct the circuit shown in Figure 58-1 for forward and reverse operation of a dc shunt motor.

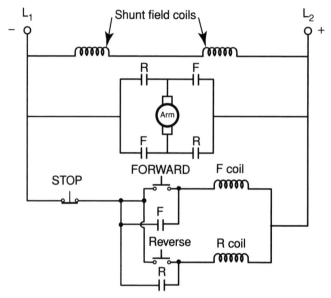

Figure 58-1. Control circuit for forward and reverse operation of a dc shunt motor.

2. Apply rated voltage to the dc shunt motor and check the circuit for operation. The FORWARD, REVERSE, and STOP switches should function. Have your instructor check the circuit for forward and reverse control.

Instructor's Approval: _____

3. Disassemble the dc shunt motor control circuit and connect the circuit shown in Figure 58-2 for forward and reverse operation of a three-phase induction motor.

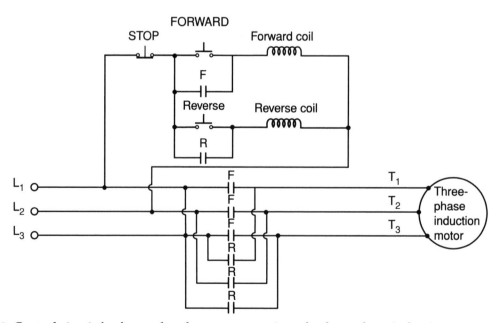

Figure 58-2. Control circuit for forward and reverse operation of a three-phase induction motor.

Activity 58—Forward and Reverse Control Circuits

4. Apply the rated voltage to the three-phase motor and check the circuit for operation. The FORWARD, REVERSE, and STOP switches should function. Have your instructor check the circuit for forward and reverse control.

Instructor's Approval: _____

Analysis

1. How is the direction of rotation of a dc shunt motor changed? _____

_____

2. How is the direction of rotation of a three-phase induction motor changed? _____

_____

3. Explain the operation of the forward/reverse control of the dc shunt motor. _____

_____

_____

4. Explain the operation of the forward/reverse control of the three-phase induction motor. ____

_____

_____

Activity 59—Forward and Reverse Control Wiring for Three-Phase Motors

Name _____ Date _____ Score _____

Objectives

In this activity, you will create control and wiring diagrams for a three-phase ac motor. You will also observe the forward and reverse operation of a three-phase ac motor.

Equipment and Materials

Three-phase ac power source

- Three-phase ac induction motor

Magnetic contactors—three-phase (2)

- Start/stop push-button station
- Conductors—#12 AWG

Safety

Be sure to wear eye protection while electric motors are in operation. Also, be very careful when working with high voltages.

Procedure

1. A control panel for forward and reverse control of a three phase ac motor is shown in Figure 59-1. This is a pictorial diagram for this laboratory activity.

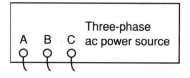

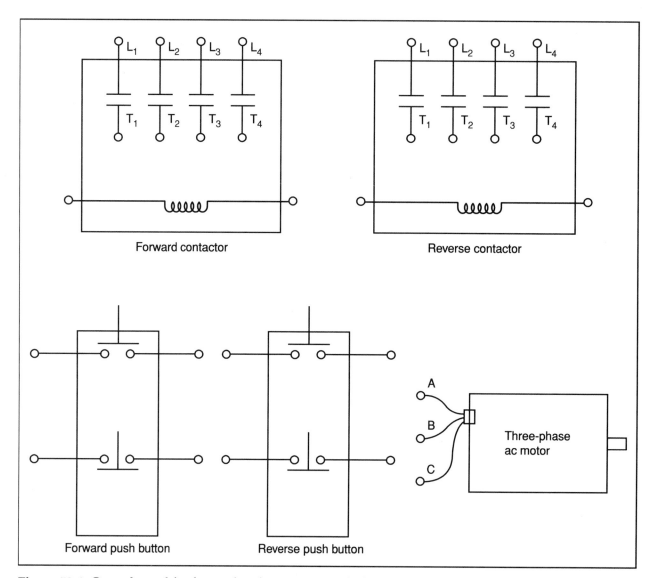

Figure 59-1. Control panel for forward and reverse control of a three-phase alternating current motor.

Activity 59—Forward and Reverse Control Wiring for Three-Phase Motors

2. Make a sketch of the *control diagram* for forward/reverse/stop control of a three-phase ac motor.

3. Complete the *wiring diagram* of the pictorial above to correspond to your control diagram. Use Figure 59-1.

4. Wire the circuit of Figure 59-1 to accomplish forward and reverse control of a three-phase ac motor.

5. Have your completed circuit checked for proper operation.

   Instructor's Approval: _____

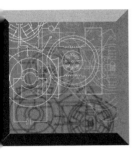

Activity 60–Forward and Reverse Control Wiring for Direct Current Motors

Name _____ Date _____ Score _____

Objectives

In this activity, you will create control and wiring diagrams for a dc shunt-wound motor. You will also observe the forward and reverse operation of a dc motor.

Equipment and Materials

- DC power source
- DC shunt-wound motor
- Magnetic contactors (2)
- Start/stop push-button station
- Conductors—#12 AWG

Safety

Be sure to wear eye protection while electric motors are in operation. Also, be very careful when working with high voltages.

Procedure

1. A control panel for forward and reverse control of a direct current shunt-wound motor is shown in Figure 60-1. This is a pictorial diagram for this laboratory activity.

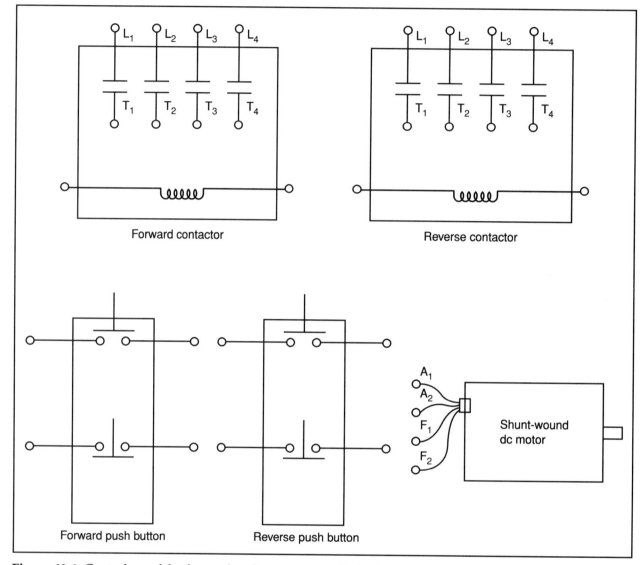

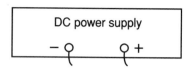

Figure 60-1. Control panel for forward and reverse control of a direct current motor.

2. Make a sketch of the *control diagrams* for forward/reverse/stop control of a dc shunt-wound motor.

3. Complete the *wiring diagram* of the pictorial to correspond to your control diagram. Use Figure 60-1.

4. Wire the circuit of Figure 60-1 to accomplish forward and reverse control of a direct current (dc) motor.

5. Have your completed circuit checked for proper operation.

   Instructor's Approval: _____

Activity 61–Dynamic Braking

Name _____ Date _____ Score _____

Objectives

When a motor is turned off, its shaft will continue to rotate for a short period of time. This continued rotation is undesirable for many industrial applications. Dynamic braking is a method used to bring a motor to a quick stop when power is turned off. Motors with wound armatures utilize a resistance connected across the armature as a dynamic braking method. When power is turned off, the resistance is connected across the armature. This causes the armature to act as a loaded generator, making the motor slow down immediately.

In this activity, you will use simple dynamic braking methods to stop a single-phase ac induction motor and a dc shunt motor.

Equipment and Materials

- Single-phase ac induction motor (120 volts)
- DC shunt motor
- Double-pole, double-throw switch
- Resistor—10 Ω, 20 W

Safety

Be sure to wear eye protection while electric motors are in operation. Also, be very careful when working with high voltages.

Procedure

1. Construct the dynamic brake for a single-phase ac induction motor as shown in Figure 61-1.

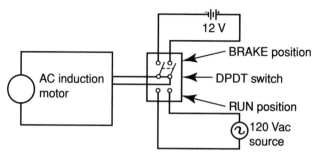

Figure 61-1. Dynamic braking circuit for a single-phase ac induction motor.

2. Place the DPDT switch in the RUN position, applying 120 volts ac to the motor.

3. Open the switch (center position) and observe the time required for the motor shaft to stop rotating.

   Estimated time for the motor to stop = _____ seconds.

4. Once again place the DPDT switch in the RUN position. Now, place the switch in the BRAKE position while observing the time required to stop the motor.

   Estimated time for the motor to stop = _____ seconds.

5. This concludes the dynamic braking procedure for a single-phase induction motor. Turn off the power and disconnect the single-phase ac induction motor.

6. Construct the circuit shown in Figure 61-2 to demonstrate dynamic braking of a dc shunt motor.

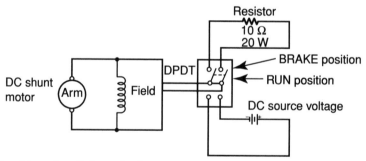

Figure 61-2. Dynamic braking circuit for a dc shunt motor.

7. Apply rated voltage to the dc motor by placing the DPDT switch in the RUN position.

8. Open the switch (center position) and observe the time required for the motor shaft to stop rotating.

   Time for motor to stop = _____ seconds.

9. Again place the switch in the RUN position. Now, place the switch in the BRAKE position while observing the time required for the motor to stop.

   Time for motor to stop = _____ seconds.

10. Turn off the power and disassemble your circuit. Return all materials.

Analysis

1. What are some situations where dynamic braking of motors is needed? _____

    _____

    _____

2. Discuss the method used for dynamic braking of an induction motor. _____

    _____

    _____

3. Discuss the method used for dynamic braking of a dc motor. _____

    _____

    _____

4. Draw a schematic diagram of a switching system that could be used for dynamic braking of a *series wound dc motor.*

Activity 62–Industrial Electronic Symbols

Name _____ Date _____ Score _____

Objectives

There are certain electronic symbols that are fundamental to industrial control systems. The majority of the symbols selected for this activity are used frequently in control circuits. You should be familiar with each symbol as a basis of understanding schematic diagrams.

In this activity, you should review the representative symbols, Figure 62-1, and demonstrate your understanding by completing the analysis.

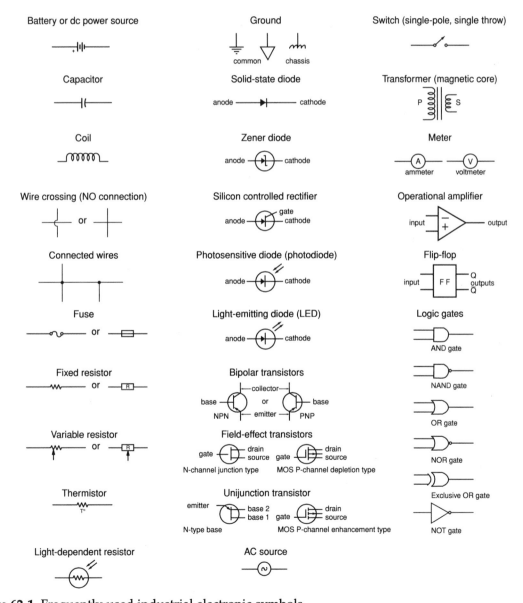

Figure 62-1. Frequently used industrial electronic symbols.

Analysis

Without looking at the symbols in Figure 62-1, identify each of the following symbols. Place the correct response in the space provided

1. _____

2. _____

3. _____

4. _____

5. _____

6. _____

7. _____

8. _____

9. _____

10. _____

11. _____

12. _____

13. _____

14. _____

15. _____

16. _____

17. _____

18. _____

Activity 63–Characteristics of Diodes

Name _____ Date _____ Score _____

Objectives

The *I-V* characteristics of a diode tell a great deal about its performance in an operating circuit. The data for an *I-V* curve can be determined by voltage and current measurements obtained from an operating circuit. This data can then be plotted on a graph and used to see how a diode will respond when it is placed in operation.

A number of important diode characteristics can be obtained from the data of an *I-V* curve. The characteristics are: the point where forward voltage causes an increase in forward current, the effect that reverse voltage has on reverse current, the influence that temperature has on forward and reverse conduction, and the conduction differences between germanium and silicon. The forward and reverse resistance of a diode can also be determined from *I-V* data. This information is very useful in predicting diode performance in a circuit.

In this activity, you will:

1. Determine the forward voltage, current, and resistance of a silicon and a germanium diode.

2. Determine the reverse voltage, current, and resistance of a silicon and a germanium diode.

3. Plot an *I-V* curve for a silicon and a germanium diode.

Equipment and Materials

- Multimeter

- Variable dc source

- Silicon diode, Texas Instruments 1N4001

- Germanium diode, General Electric 1N60

- Resistor—1 kΩ, 1/4 W

- Circuit construction board

Procedure

1. Construct the circuit shown in Figure 63-1. The multimeter is used to measure different circuit voltages. It must be moved to different locations to observe these voltages.

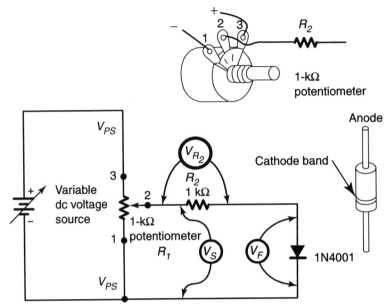

Figure 63-1. Diode characteristic testing circuit.

2. Turn on the dc voltage source (V_{PS}) and adjust it to produce 10 V, as measured across the 1-kΩ potentiometer. This voltage will appear across the two outside terminals (1 and 3) of the potentiometer. The source voltage (V_S) for the circuit is a variable value that is adjusted by the potentiometer. V_S will appear at terminals 1 and 2 of the potentiometer. In this activity, you are directed to adjust potentiometer R_1 to produce a specific voltage across the diode (V_F). Then measure the voltage across the series resistor (R_2) and circuit source voltage across terminals 1 and 2 of the potentiometer. Record these values in Data table 1 of Figure 63-2.

3. The first part of this experiment is designed to measure the forward *I-V* characteristics of a diode. Note that only voltage values are called for in the procedure. Forward current (I_F) is determined by calculation. Ohm's law is used to determine current by the formula $I_F = V_{R_2}/R_2$. Forward resistance R_F is determined from the expression $R_F = V_F/I_F$.

4. Adjust V_S to produce a V_F of 0 V across the diode. Measure and record the values of V_S and V_{R_2}.

5. Adjust the value of V_S to produce a V_F of 0.1 V across the diode. Measure the values of V_S and V_{R_2}. Record the measured values of V_S and V_{S_2} in Data table 1.

6. Repeat step 3 for each of the V_F values of Data table 1. Note that the V_F voltage of a silicon diode is generally considered to be 0.7 V. In an operating circuit, this may vary from 0.5 to 0.7 V according to the amount of current flowing through the diode. As a rule, a large V_F will appear across a diode only when the value of I_F has been increased significantly. In this circuit, the value of I_F is kept at a rather low value. With some diodes, the upper value of V_F may not be obtained. If this occurs, use only the value that can be achieved by the circuit. Do not try to reach the upper values by increasing the power source voltage.

7. Calculate the I_F and R_F values for each of the measured V_F values.

1N4001				
Data table 1				
Measured			Calculated	
V_F	V_{R_2}	V_S	I_F	R_F
0				
0.1				
0.2				
0.3				
0.4				
0.5				
0.6				
0.7				

1N4001				
Data table 2				
Measured			Calculated	
V_R	V_{R_2}	V_S	I_F	R_F
0				
2				
5				
7				
10				
12				
15				
20				

1N60				
Data table 3				
Measured			Calculated	
V_F	V_{R_2}	V_S	I_F	R_F
0				
0.1				
0.2				
0.3				
0.4				

1N60				
Data table 4				
Measured			Calculated	
V_R	V_{R_2}	V_S	I_F	R_F
0				
2				
5				
7				
10				
12				
15				
20				

Figure 63-2. Data tables.

8. Turn off the power source and remove the 1N4001 diode from the circuit. Reverse the two leads and return it to the circuit so that it is connected in the reverse-bias direction. The meter will be used to measure reverse voltage values across the diode (V_R), resistor voltage (V_{R_2}), and the voltage source (V_S). The reverse current (I_R) and reverse resistance (R_R) are calculated from these values.

9. Turn on the variable dc power source and adjust it to 20 V measured across terminals 1 and 3 of the potentiometer. Repeat steps 2 through 5 for the reverse-connected diode. The values of V_R are indicated in Data table 2. Note that these values are much larger than the previous V_F values.

10. Turn off the power source.

11. Using the V_F and I_F data from Data table 1, plot a forward conduction curve on the graph of Figure 63-3. The forward curve is located in the upper right quadrant of the graph.

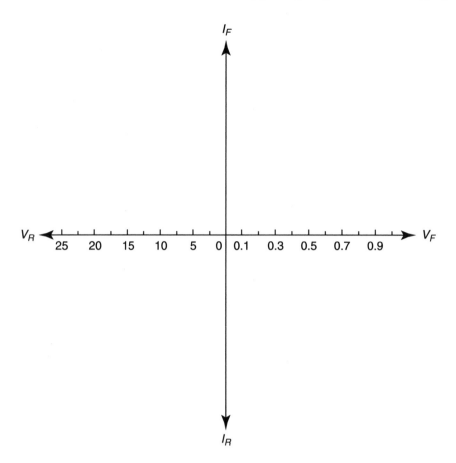

Figure 63-3. Chart of *I-V* characteristics.

12. Using the V_R and I_R data from Data table 2, plot a reverse conduction curve on the graph of Figure 63-3. The reverse curve is located in the lower left quadrant of the graph.

13. Repeat step 1 through 12 for the 1N60 germanium diode. In this procedure, the V_F of the 1N60 diode should not be adjusted to values in excess of 0.4 V. Fill in the appropriate measurements and calculations into Data tables 3 and 4. When plotting the forward and the reverse curves on the graph of Figure 63-3, use dots or dashed lines to distinguish 1N60 data from the 1N4001.

14. Turn off the power to your circuit. Return all materials.

Analysis

1. Explain what the forward characteristics of a silicon diode tell about its operation. _____

_____

_____

2. Explain what the reverse characteristics of a silicon diode tell about its operation. _____

_____

_____

Activity 63—Characteristics of Diodes

Activity 64–Zener Diode Testing

Name _____ Date _____ Score _____

Objectives

An ohmmeter can be used to identify the leads of a zener diode and to detect whether it is shorted or open. The ohmmeter, however, does not determine how a zener diode will respond when voltage is applied. A simple test procedure can be performed to evaluate this condition of operation using a variable dc source, a series resistor, and a zener diode. Its zener voltage can be determined by altering the variable source voltage while observing the voltage across the zener. The voltage will remain constant across the zener when the V_Z value is reached. This will show the tolerance, the V_Z, and the knee voltage of the device. This is a significant characteristic when the zener is reverse biased.

In this activity, you will:

1. Demonstrate a practical method of testing a zener diode with a voltage source.

2. Evaluate the V_Z of a zener diode.

3. Connect a zener diode in a circuit for normal operation.

Equipment and Materials

Multimeter

Variable dc source

Zener diode—5.1 V ± 10%, 500 mW (Motorola 1N5231, 1N751 or equivalent)

Resistor—100 Ω, 1/4 W, 5% tolerance

Potentiometer—1 kΩ, 1 W

Circuit construction board

Procedure

1. Prepare the multimeter to measure resistance on its R × 1 or lowest resistance range. For a digital meter, use the 2 kΩ or an equivalent range.

2. Connect the two leads of the meter across the diode. A good zener will show low resistance when it is forward biased and infinite resistance when reverse biased. Identify the anode and cathode of the diode with the meter.

3. Make a sketch of the identified leads of the device.

4. Construct the circuit shown in Figure 64-1. The 1-kΩ potentiometer is adjusted to produce different values of input voltage. Adjust the dc power source to produce 12 V. Connect this voltage across the potentiometer as indicated.

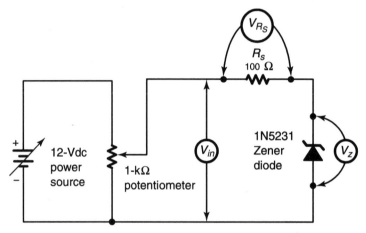

Figure 64-1. Zener diode test circuit.

5. Adjust the potentiometer to produce 0 V of input voltage for the circuit. This voltage is labeled dc input in the chart of Figure 64-2.

6. Measure the voltage that appears across the zener diode. Record this value in the chart.

7. Adjust the potentiometer to produce 1 V of input voltage. Measure the V_Z and record its value in the chart.

8. Repeat this procedure for each of the indicated input voltage values.

9. The voltage across the series resistor (V_{R_S}) of the circuit is the difference between the dc input voltage and V_Z. For a few of the input voltage values around 5 V, calculate the V_{R_S}. These voltage values can also be measured with the multimeter for the appropriate input voltage value.

10. The zener current of the circuit can be calculated using the expression $I_Z = V_{R_S}/R_S$. Calculate a few of the I_Z values of the diode around the 5 V_Z part of the chart.

11. Reverse the zener diode in the circuit of Figure 64-1. Adjust the potentiometer to produce an input voltage of 0 V. Measure and record the diode voltage V_F and the series resistor voltage V_{R_S}. Adjust the input voltage across the potentiometer to 0.1 V. Again measure and record the V_F and V_{R_S}.

DC V_{in}	V_Z	V_{R_S}	I_Z
0			
1			
2			
3			
4			
5			
6			
7			
8			
9			
10			
11			
12			

DC V_{in}	V_F	V_{R_S}	I_F
0			
0.1			
0.2			
0.3			
0.4			
0.5			
0.6			
0.7			
0.8			
0.9			
1.0			

Figure 64-2. Data tables.

12. Repeat the procedure for each of the dc input values in the chart on the right side of Figure 64-2. Note that the input voltage does not exceed a value in excess of 1 V. This diode conducts very heavily in the forward direction. This should be noted by the observed voltage values. An increase in the value of V_F indicated a corresponding increase in I_F.

13. Calculate the forward current for the zener diode for the measured values of the chart. $I_F = V_{R_S}/R_S$. Record the calculated values in the chart.

14. Turn off the power source and disconnect the circuit. Return all materials.

Analysis

1. Explain what causes a zener diode to maintain its V_Z at a constant value when the source voltage is increased. _____

_____

_____

2. Explain why the power dissipation rating of a zener diode is an important selection characteristic.

_____

_____

_____

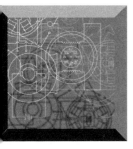

Activity 65—Characteristics of Transistors

Name _____ Date _____ Score _____

Objectives

Any material can be electrically classified as a conductor, semiconductor, or insulator. Conductors are materials, such as copper or aluminum, through which an electric current will easily flow. Insulators are materials like rubber and glass, through which an electric current will not easily flow. Semiconductors are materials or devices classified somewhere between conductors and insulators because current through them can be controlled. This is achieved through controlling the resistance of semiconductors by using a bias current. When a semiconductor is allowed to conduct at its maximum, it exhibits the characteristics of a conductor. When a semiconductor is biased to not allow conduction it exhibits the characteristics of an insulator. A semiconductor can be controlled to allow conduction between these two extremes.

The most common of all semiconductor devices is the transistor. Transistors are classified as NPN or PNP. These devices have three external leads: emitter, collector, and base. Figure 65-1 shows the electrical symbols, materials, and lead placement of the transistor.

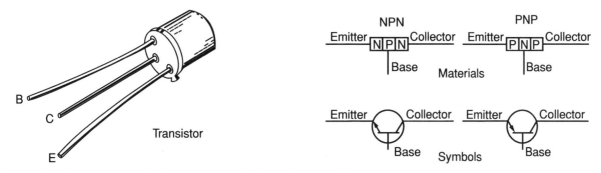

Figure 65-1. Transistor lead placement, symbols, and materials.

Transistors are used as ac or dc amplifiers and can be connected into circuits in one of three configurations: common-emitter, common-collector, or common-base. These three configurations are shown in Figure 65-2.

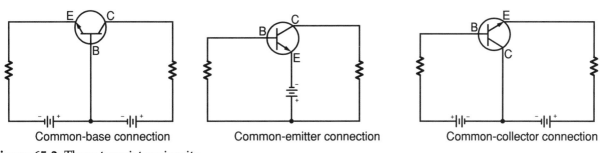

Figure 65-2. Three transistor circuits.

In this activity you will observe the dc characteristics of an NPN transistor while it is connected in the common-emitter configuration.

Equipment and Materials

- Multimeter
- Variable dc power supply
- SPST switches (2)
- NPN transistor—2N2405
- Resistors—200 Ω, 27 kΩ , 33 kΩ, 39 kΩ, 51 kΩ, 68 kΩ, 100 kΩ, 200 kΩ
- Connecting wires

Procedure

1. With the multimeter adjusted to measure resistance in the R × 1 ohms range, measure and record the resistance between the leads of the 2N2405 NPN transistor. *Note:* Be sure to observe the meter polarity as indicated in the following illustrations.

   a. When the emitter is (–) and the base is (+), their forward resistance = _____ Ω.

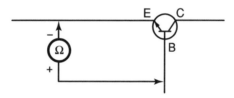

   b. When the emitter is (+) and the base is (–), their reverse resistance = _____ Ω.

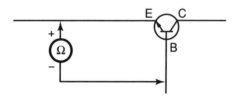

   c. When the collector is (–) and the base is (+), their forward resistance = _____ Ω.

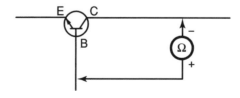

   d. When the collector is (+) and the base is (–), their reverse resistance = _____ Ω.

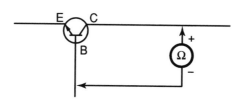

e. When the emitter is (–) and the collector is (+), their resistance = _____ Ω.

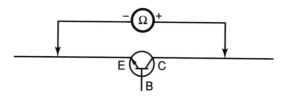

f. When the emitter is (+) and the collector is (–), their resistance = _____ Ω.

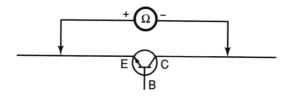

2. How did the forward resistance of the emitter–base in Step 1a compare with the reverse resistance in Step 1b?

_____

_____

3. How did the forward resistance of the collector–base in Step 1c compare with the reverse resistance in Step 1d?

_____

_____

4. How did the resistance of the emitter–collector in Step 1e compare with the resistance in Step 1f?

_____

_____

5. If the resistances of an identical PNP transistor had been measured, how would these compare with the resistances of the NPN transistor?

_____

_____

6. How would the ohmmeter polarity differ when measuring the resistance of the PNP transistor?

_____

_____

7. Construct the circuit in Figure 65-3. *Note:* The meter should be adjusted to measure direct current in the 100-mA range.

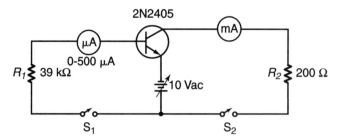

Figure 65-3. Common-emitter circuit.

8. Allow S_2 to remain open, close S_1, and record the emitter–base current.

    E-B current = _____ µA

9. Open S_1, close S_2, and record the emitter–collector current.

    E-C current = _____ mA

10. Close both S_1 and S_2. Record both the emitter–base and emitter–collector currents in the space below.

    E-B current = _____ µA

    E-C current = _____ mA

11. How did the E–B and E–C currents in Steps 8 and 9 compare with those in Step 10?

    _____

    _____

12. With both S_1 and S_2 closed, connect the proper values of R_1 into the circuit illustrated in Step 7 and complete Figure 65-4.

Value of R_1	Measured base current	Measured collector current	Computed emitter current	Computed R_2 voltage	Computed emitter-collector voltage
27 kΩ					
33 kΩ					
39 kΩ					
51 kΩ					
68 kΩ					
100 kΩ					
200 kΩ					

Figure 65-4. Effects of changing R_1.

13. Using the data from the chart completed in Step 12, complete the following statements:
   a. As the value of R_1 increases, the base current _____.
   b. As the base current decreases, the collector current _____.
   c. As the base current decreases, the emitter current _____.
   d. The _____ current controls the amount of _____ and _____ current.
   e. As the base current decreases, the voltage across R_2 _____.
   f. As the base current decreases, the emitter–collector voltage _____.
   g. As the voltage across R_2 decreases, the emitter–collector voltage _____.
   h. The sum of the base and collector currents equals the _____ current.
   i. The _____ current controls the _____ current, which controls the _____ and _____ voltages.
14. Why is the transistor used as an amplifier?

_____

_____

15. Turn off the power and disassemble your circuit. Return all materials.

Analysis

1. What is a semiconductor?

_____

_____

2. What are two classifications of transistors?

_____

_____

3. In the space below, draw the electrical symbols for the PNP and NPN transistors.

4. What are the three external leads of a transistor?

_____

_____

5. How do the forward and reverse resistances of the emitter–base areas compare?

_____

_____

6. What three circuit configurations are most commonly associated with transistors?

_____

_____

7. What is the relation between base current and collector current when a transistor is connected in the common–emitter configuration?

_____

_____

8. What is the relation between base current and collector–emitter voltage when a transistor is connected in the common-emitter configuration?

_____

_____

Activity 66–Bipolar Transistor Testing

Name _____ Date _____ Score _____

Objectives

The internal circuitry of an ohmmeter has a source voltage connected to its leads when measuring resistance. This voltage can be used to bias a transistor when it is connected to the ohmmeter. A bipolar transistor has two PN or NP junctions in its construction. When these junctions are connected together, an NPN or PNP type of transistor is formed. A very important step in transistor testing is evaluation of the two junctions. They must respond like diodes in order for a transistor to be good. This means that forward biasing will cause low resistance, and reverse biasing will show infinite or high resistance. The two outside leads of a transistor will show infinite or high resistance to either method of biasing.

A good bipolar transistor must show low resistance when its emitter–base and base–collector are forward biased. Reversing the polarity of the ohmmeter will cause the same two junctions to have an extremely high resistance. The emitter–collector junctions must show high resistance to either ohmmeter polarity if the device is good. If it does not, the transistor has leakage current. Silicon transistors generally have no leakage, while germanium transistors show some leakage. Leakage current is temperature dependent in bipolar transistors.

In this activity, you will:

1. Test the status of a bipolar transistor with an ohmmeter.

2. Identify the polarity of the construction material of a bipolar transistor with an ohmmeter.

3. Identify the leads of a bipolar transistor with an ohmmeter.

4. Test the gain capabilities of a bipolar transistor with an ohmmeter.

Equipment and Materials

- Multimeter
- Transistor (Motorola MPSA20 or equivalent)
- Transistor (Motorola MPSA70, 2N2431 or equivalent)
- Resistor—10 kΩ, 1/4 W

Procedure

1. Determine the lead polarity of the meter by placing it in the ohm position and measuring the voltage using a second voltmeter. If a second voltmeter is not available, determine the polarity of the meter from the instrument operational manual.

   The polarity of the meter is (straight, reversed).

265

2. In the first testing operation, we will determine if the transistor is NPN or PNP. The polarity of the meter is very important in this procedure. Set the range switch of the meter to the R × 100 or R × 1000 position. If a digital meter is used set the range–select switch to the 2 kΩ position or an equivalent range.

3. Select the MPSA70 transistor for this test procedure. Look at the three leads of the transistor. Notice if there are any identification marks or unusual lead placement for this transistor. Make a sketch of the lead placement of the transistor in the outer margin. Show the bottom view.

4. Pick up the center lead of the transistor and assume that it is the base. Connect the negative lead of the meter to it. If the lead is actually the base of the transistor, it will show a low resistance reading between each of the other two leads for a PNP transistor.

5. If the resistance in Step 4 is high in both directions, reverse the meter leads so that the positive lead goes to the assumed base. Once again test the transistor by connecting the negative lead alternately between the two outermost leads. If a low resistance reading is obtained between the center and outside leads, the transistor is an NPN type.

6. If using the center lead does not produce a low resistance between the two outside leads, it is not the base terminal. Select one of the outside leads and assume that it is the base. Try the same test procedure again. Note that a shorted transistor will more than likely show low resistance in either polarity direction. It is rather difficult to determine the polarity of a shorted transistor.

7. After you have determined the polarity of a transistor to be either NPN or PNP by this procedure, determine if the transistor has any leakage. This test is applied to the two outside leads or to the two leads that do not include the assumed base. These leads will be identified as the emitter and collector. In general, there should be no leakage current or resistance in either direction that the meter is connected. Test the emitter to collector for leakage. Record the leakage resistance.

   Leakage resistance is _____ Ω.

8. We have assigned the names emitter, base, and collector to the leads of the transistor. We must now verify if the lead assignment is correct. To perform this test, assume that the base has been identified and that the transistor is of the PNP type. Select one lead and assume that it is the emitter and that the other lead is the collector. Connect the positive lead of the meter to the assumed emitter and the negative lead to the assumed collector. Then connect a 10-kΩ resistor between the collector and the base. If the lead assignments are correct, the meter will show a substantial change in resistance. This is an indication of transistor gain or beta. If there is no change in resistance, the lead assignments for the emitter and collector have been reversed. Switch the meter leads and again connect the resistor between the newly assigned collector (negative meter lead) and the base. This should produce a change in resistance indicating current gain or beta. If the meter responds in this manner, the lead assignments of the emitter, base, and collector are correct.

9. Make a bottom-view drawing of the transistor in the outer margin and identify the location of the leads.

10. Repeat the test procedure for the MPSA20 transistor. The drawing, type of transistor, and lead identification should be placed in the space that follows.

11. Turn off your meter and return all materials.

Analysis

1. Explain how an ohmmeter can be used to determine if a transistor is an NPN or a PNP type.

_____

_____

2. How is the leakage of a transistor evaluated with an ohmmeter? _____

_____

_____

Activity 67–Field-Effect Transistor Testing

Name _____ Date _____ Score _____

Objectives

The internal circuitry of an ohmmeter has a voltage source connected to its lead when it is used to measure resistance. This voltage can be used to bias a field-effect transistor. The status of the terminals, the condition of the device as good or bad, and to some extent the leads can be identified with an ohmmeter. An ohmmeter is a quick and easy method of field-effect transistor evaluation. This method of evaluation is rather commonly used in electronic troubleshooting procedures.

In this activity, you will:

1. Test the status of a field-effect transistor with an ohmmeter.

2. Identify the polarity of the construction materials of a field-effect transistor.

3. Identify the leads of a field-effect transistor.

Equipment and Materials

- Multimeter

- N-channel JFET (Motorola 2N5459 or equivalent)

- N-channel MOSFET (Motorola 2N3796 or equivalent)

- Resistor—100 kΩ, 1/4 W

Procedure

1. Determine the lead polarity of the meter by placing it in the ohms position and measuring the probe voltage with a voltmeter. Measure the voltage on the R × 1000, 2 kΩ, or equivalent range. If the common lead is negative and the probe lead is positive, the meter has forward polarity. An opposite polarity indicates reverse meter polarity.

   Record the voltage of the meter: _____ V.

   The meter used for this test has (forward, reverse) polarity.

2. Select the 2N5459 JFET. Make a sketch of the bottom view of the transistor lead layout in the right-hand margin.

3. First determine the source (S) and the drain (D) leads of the transistor by connecting the multimeter, set to measure resistance, between two randomly selected leads. If the two selected leads indicate a resistance, reverse the meter probes. Two leads that show the same resistance in either direction of polarity are the source and drain. It may be necessary to try several different two-lead combinations in order to determine the source and drain of the JFET. When the leads have been identified, indicate your findings on the JFET sketch of Step 2.

   The measured source-drain resistance is _____ Ω.

4. The gate-channel of a JFET responds as a diode. This can be observed by connecting the meter between two leads and then reversing the meter leads. The gate–source or gate–drain will respond as a diode by showing low forward resistance and high or infinite reverse resistance.

5. If the gate shows low resistance when the positive probe is connected to the gate and the negative to either the source or drain, the JFET is an N-channel device. A P-channel device will produce low resistance when the gate is negative and the source or drain is positive. The resistance ratio between forward and reverse connections should be 1:100 for a device to be good.

6. An approximate test of the JFET's ability to amplify is displayed by connecting the meter to the source and drain. The polarity of the meter is not significant. Touch the gate of the JFET with your finger. It should cause a decrease in resistance.

7. A more significant resistance change occurs when a 100-kΩ resistor is connected between the gate and either the source or drain. Very little or no change in resistance during this test indicates an open device or improper lead identification.

8. Using the same procedure outlined in Steps 2 through 6, test the 2N3796 MOSFET. This device has a fourth lead that is connected to the substrate. The substrate and the source or drain leads will respond as a diode. The gate of a MOSFET will show an infinite resistance between all leads in either direction of meter polarity.

9. Turn off your meter and return all materials.

Analysis

1. Explain how an ohmmeter can be used to evaluate the status of a JFET. _____
   _____
   _____

2. How would an ohmmeter test distinguish between an N-channel and a P-channel JFET? ____
   _____
   _____

3. What are some of the differences that an ohmmeter will show between a JFET and a MOSFET?
   _____
   _____
   _____

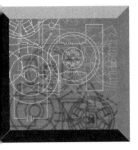

Activity 68–MOSFET Characteristics

Name _____ Date _____ Score _____

Objectives

In this experiment, you will develop a partial I_D–V_{DS} family of characteristic curves for a MOSFET. The curves are derived by setting the gate voltage to a specific value and adjusting the source–drain voltage to specific values. The drain resistor voltage is then measured for each V_{DS} value and used to calculate the drain current. The V_{DS} and I_D values are then used to plot an I_D–V_{DS} curve for the MOSFET. Several other V_{GS} values are used to plot representative curves. The polarity of the gate voltage is then reversed to show how the MOSFET will respond with a different polarity of gate voltage. This curve is also plotted.

The developed family of I_D–V_{DS} curves of the MOSFET are then used to show the operational characteristics of the device in a representative circuit. A load line is plotted for the circuit. An operating point is established according to the circuit's data, and characteristic values of drain current and source–drain voltage are determined from the curves. These values are used to calculate transconductance and voltage gain. This procedure shows how the amplifier circuit responds graphically.

In this activity, you will:

1. Demonstrate the operating characteristics of a MOSFET by measuring some representative source–drain voltages and calculating the drain current.

2. Plot a partial family of I_D–V_{DS} characteristic curves for a MOSFET.

3. Determine a load line for a representative circuit and plot it on the partial characteristic curves.

4. Evaluate the operation of a MOSFET circuit with a family of I_D–V_{DS} curves and determine voltage gain, transconductance, and amplifier phase relationships.

Equipment and Materials

- Multimeter
- Dual-section variable power supply—±15 V (or two 9-V batteries)
- N-channel MOSFET (Motorola 2N3796 or equivalent)
- Resistor—100 Ω, 1/4 W
- Resistor—1 kΩ, 1/4 W
- Resistor—4.7 kΩ, 1/4 W
- Potentiometer—1 kΩ, 1 W (linear taper)
- Potentiometer—100 kΩ, 1 W (linear taper)
- Circuit construction board

Procedure

Part A: Characteristic Curve Data Collection

1. Construct the MOSFET characteristic curve test circuit of Figure 68-1. The leads or pin-out of the MOSFET are shown in the diagram.

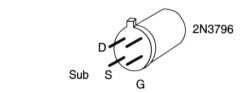

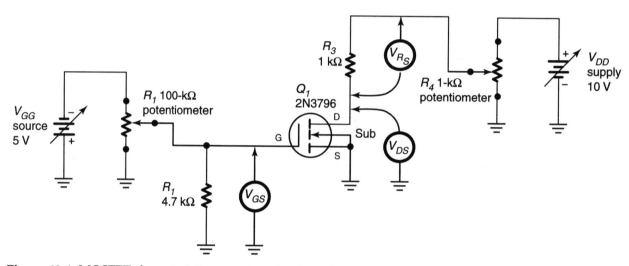

Figure 68-1. MOSFET characteristic curve test circuit.

2. With circuit construction complete, remove the MOSFET shorting ring or conducting foam.

3. Adjust the two power supply voltage controls to the 0 V position and turn on the power supply. Adjust the V_{DD} supply to produce 10 V. This is measured across the two outside leads of potentiometer R_4. Adjust the V_{GG} supply to produce 5 V. This is measured across the two outside leads of potentiometer R_1. Potentiometers R_1 and R_4 are used to change the value of the gate–source voltage and the source–drain voltage of the MOSFET.

4. Voltage measured across the drain resistor (R_3) is used to determine the drain current by calculation. $I_D = V_{R_3}/R_3$. This calculation will be made for each V_{DS} voltage value and recorded in the data column of the chart as I_D.

5. V_{GS} is measured across the gate–source resistor R_2. This voltage is adjusted to a specified value and then V_{DS} is adjusted to the values indicated in the chart of Figure 68-2.

	$V_{GS} = 0$ V		$V_{GS} = -1$ V		$V_{GS} = +1$ V	
V_{DS}	Measured V_{R_3}	Calculated I_D	Measured V_{R_3}	Calculated I_D	Measured V_{R_3}	Calculated I_D
0 V						
1 V						
2 V						
4 V						
6 V						
8 V						
10 V						

Figure 68-2. Chart of I_D–V_{DS} values.

6. Connect the meter across the gate–source of the MOSFET and adjust R_1 to produce 0 V_{GS}. Connect the meter across the source–drain and adjust the potentiometer R_4 to produce 0 V_{DS}. Measure and record the voltage across resistor R_3. Calculate the drain current for this value. Adjust potentiometer R_4 to produce a V_{DS} of 1 V. Measure and record the voltage across resistor R_3 for this V_{DS} setting. Calculate the drain current and record its value in the chart. Repeat this procedure for V_{DS} values of 2, 4, 6, 8, and 10 V.

7. Return the value of V_{DS} to 0 V.

8. Increase the value of V_{GS} to –1 V. Repeat the procedure outlined in Step 6. Record your data values in the –1 V_{GS} column of the chart.

9. Return the two potentiometers to 0 V. Disconnect the negative lead of the power supply connected to potentiometer R_1. Connect this potentiometer lead to the positive side of the V_{DD} power supply. Adjust the V_{DD} power supply to 10 V.

10. Adjust the potentiometer R_1 to produce +1 V_{GS}. Repeat the procedure outlined in Step 6 for different values of V_{DS} for 0 to 10 V. Record you data in the +1 V_{GS} column of the chart of Figure 68-2.

11. Using the data collected in the chart of Figure 68-2, plot a partial set of I_D–V_{DS} curves for the MOSFET in Figure 68-3.

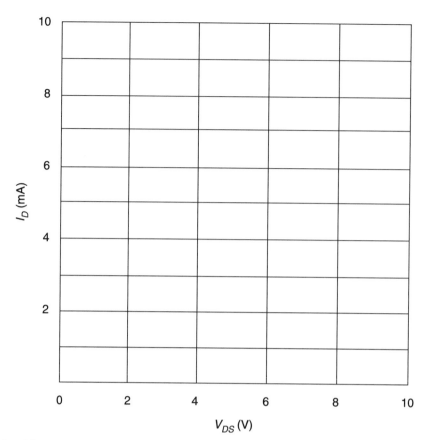

Figure 68-3. Graph of I_D versus V_{DS}.

Part B: Load Line Analysis

1. Plot a load line on the partial set of I_D–V_{DS} curves of Figure 68-3 for the circuit of Figure 68-4.

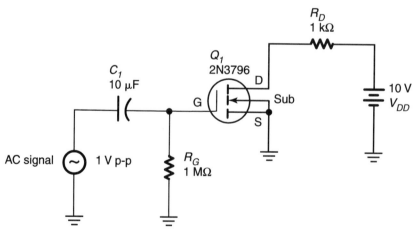

Figure 68-4. MOSFET circuit for load line analysis.

Activity 68—MOSFET Characteristics

2. Resistor R_D serves as the load, and V_{DD} serves as the source. The two conditions of operation are full conduction and nonconduction.

3. The operating point of this circuit is zero, based on the value of V_{GS}. Mark the Q point on the load line.

4. Project lines from the Q point to the I_D scale and to the V_{DS} scale of the graph.

   $I_D =$ _____ mA; $V_{DS} =$ _____ V.

5. With the indicated ac signal, the total change in V_{GS} is _____ peak to peak. The resulting output from this input is _____ V p-p. Note the polarity of V_{DS} compared with V_{GS}.

   What does this indicate? _____.

6. The resulting I_D that occurs for this input is _____ mA p-p.

   How does the polarity of I_D compare with V_{DS}? _____.

7. Using the value of the indicated input and the output voltage, calculate the voltage gain (A_V) for the circuit.

   $A_V =$ _____.

8. Transconductance is a measure of the effectiveness of this device to amplify a signal. Transconductance $= I_D/V_{GS}$. Determine the transconductance of the amplifier circuit using your data.

   Transconductance = _____ S.

9. Turn off the power and disassemble your circuit. Return all materials.

Analysis

1. Explain how an N-channel E-type MOSFET achieves amplification. _____

   _____

   _____

2. How would an increase in the resistance of the drain resistor or load change the operation of a MOSFET? _____

   _____

   _____

3. What are the operational differences between an E-type MOSFET and a JFET amplifier? _____

   _____

   _____

Activity 69–SCR Testing

Name _____ Date _____ Score _____

Objectives

The internal circuitry of an ohmmeter has a voltage source connected to its leads when it is used to measure resistance. This voltage can be used to bias the junctions of an SCR. The good/bad status of the device and its leads can be identified through this process. The ohmmeter is a quick and easy method of evaluating an SCR. This method of evaluation is used widely in electronic troubleshooting procedures.

As a rule, an analog multimeter or hand-deflection meter works very well for this method of evaluation. A digital multimeter does not work as well for some of the measurements. Digital instruments generally have a very small current flow when measuring resistance. This can restrict the test procedure by not permitting the meter to trigger an SCR and have it latch or hold in its conduction range. All other test procedures are essentially the same.

In this activity, you will:

1. Test the status of a silicon-controlled rectifier with an ohmmeter.

2. Identify the leads of a silicon-controlled rectifier.

Equipment and Materials

- Multimeter

- Silicon-controlled rectifier (General Electric C106F or equivalent)

- Circuit construction board

Procedure

1. Prepare the multimeter to measure resistance in the low resistance range. Use R × 1 for analog meters and 2 kΩ or equivalent for a digital instrument.

2. Select any two SCR leads at random and test the forward and reverse resistance ratio. A two-lead combination that responds as a diode (low-resistance in one direction and infinite resistance in the other direction) is the cathode–gate. A two-lead combination that shows infinite or high resistance in both directions is either the anode–cathode or the gate–anode. Find the two-lead combination that responds as a diode. The third or alternate lead of this combination is the anode. For the two-lead diode combination, connect the meter to produce a low resistant reading. This will show the cathode connected to the negative and the gate to the positive lead. (This assumes that your meter has straight polarity.)

3. Make a sketch of the SCR in the right margin indicating the lead assignments you have selected.

4. To verify the correctness of the lead assignments, connect the meter as shown in Figure 69-1. This should initially make an infinite resistance indication on the meter.

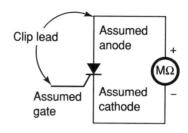

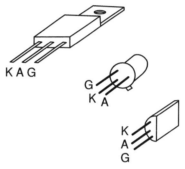

Figure 69-1. Lead assignments for testing an SCR.

5. Connect a wire between the anode and the gate. If the lead assignment is correct, the meter will change to a low resistance value. If the wire is disconnected, the meter should continue to show low resistance. This shows the latching or holding function of the SCR. If a digital meter is used, the SCR may not latch or hold in its conduction state. A standard analog meter will have enough current in the ohmmeter range to latch and hold the SCR into conduction. To alter the latching function, momentarily disconnect one side of the meter, then return the meter leads to the anode–cathode. This should show an infinite resistance with the gate lead disconnected. Connect the gate lead again to the anode to see the latching function. Try this test procedure several times to be certain of how it is performed.

6. This test procedure applies only to good SCRs. If a two-lead combination cannot be found that responds as a diode, the gate–cathode junction is open. If any two-lead combination shows low resistance in either direction of polarity, the junction or combined junctions are leaky or shorted. SCRs can become open or shorted in normal usage. As a rule, the SCR is rather free of trouble and responds as a very reliable device.

7. Turn off your meter and return all materials.

Analysis

1. Why does an analog multimeter work best in evaluating the status of an SCR? _____

   _____

   _____

2. Why does the gate–cathode of an SCR show low resistance in one direction and high resistance in the other direction of ohmmeter polarity? _____

   _____

   _____

3. Explain why the anode–cathode of a good SCR shows infinite resistance in either direction of ohmmeter polarity. _____

   _____

   _____

Activity 70–Triac Testing

Name _____ Date _____ Score _____

Objectives

The internal circuitry of an ohmmeter has a voltage source connected to its leads when it is used to measure resistance. This voltage source can be used to bias a triac into conduction. The good/bad status of a triac can be tested, and the leads can be identified. An ohmmeter is a quick and easy method of evaluating the status of a triac.

An analog ohmmeter works best for this method of evaluation. This type of instrument generally has a larger value of current in its ohmmeter. The current value is somewhat critical when triggering a triac into conduction and getting it to latch. Digital meters, as a rule, have a very nominal current flow when the meter is used as an ohmmeter. This current value is generally not large enough to cause a triac to latch. All other tests being performed will respond equally with either type of meter.

The test procedure of a triac is primarily the same as that of an equivalent SCR. The triac, however, can be triggered in four different modes or conditions. Each lead has a dual-crystal polarity that is voltage selective. The polarity of the ohmmeter is used to select the appropriate combination of the test. Identification of the respective leads is somewhat more complex than the procedure used in evaluation of the SCR.

In this activity, you will:

1. Test the status of a triac with an ohmmeter.

2. Identify the leads of a triac with an ohmmeter.

Equipment and Materials

Multimeter

Triac (RCA 2N5724 or equivalent)

Circuit construction board

Procedure

1. If an analog meter is being used, prepare it to measure ohms in the R × 1 range. If a digital meter is used, prepare it to measure resistance in the 2-kΩ or equivalent range. The R × 1 is preferred because it usually provides the necessary holding current to latch this device into conduction. We will assume that the instrument being used has straight polarity with the negative connected to the black or common lead and positive connected to the red or ohms lead.

2. Select any two of the three leads at random and determine the forward to reverse resistance ratio. Terminal 1 (T_1) to terminal 2 (T_2) should indicate an infinite resistance or extremely high resistance in both directions. The gate (G) and terminal 1 will show a rather low resistance in both directions. With the two leads that show low resistance in either direction identified, the third lead is terminal 2.

3. Make a sketch of the triac leads in the right-hand margin. Identify T_2 in the sketch. Leads T_1 and the gate have not been identified at this time.

4. The next test procedure is based on a right or wrong selection opportunity of determining the correct lead assignment for T_1 and G. Connect the positive side of the meter to T_2, then select one of the two remaining leads and connect the negative side of the meter to it. We will assume in this case that the negative meter lead is connected to T_1. Now attach a connecting wire to the third lead, which we have assumed to be the gate. Connect the other end of this wire to T_2. If the assumed T_1 and G are correct, the meter will show a low resistance, which indicates triggering. Momentarily disconnecting the gate from T_2 should cause the triac to stay in its conduction state. If this procedure does not produce a change in resistance the assumed T_1 and G are reversed. Switch the negative meter lead from the first assumed T_1 lead to the assumed gate lead. Place the connecting wire to the newly assumed gate lead and try the triggering procedure again. If triggering is produced, the correct lead assignment has been made. Label these two leads as G and T_1. If the triac cannot be triggered by this procedure it may be faulty. Only a good triac will respond to this test.

5. With the lead assignments made, once again try the triggering procedure. Connect the negative lead to T_1 and the positive lead to T_2. Then attach a connecting wire between the gate and T_2. This should cause a resistance change in the meter and trigger the triac into conduction.

6. If the lead assignments are correct, the triac should also trigger in the reverse polarity direction. In this case, connect the negative side of the meter to T_2 and the positive side to T_1. Attach a connecting wire to the gate and momentarily connect it to T_2. This action should also trigger the triac into conduction. If the triac is functioning properly, it can be triggered in either direction of polarity. Test the triac to see if it has this capability.

7. A digital meter will work for all test procedures except the latching function. The analog meter will latch the triac only when it is in the $R \times 1$ range of operation.

8. Turn off your meter and return all materials.

Analysis

1. What are some of the common things to look for in testing an SCR to see if it is faulty? _____
   _____
   _____

2. Why does the gate and T_1 of a triac show low resistance in either direction of ohmmeter polarity? _____
   _____
   _____

3. Why does T_1 and T_2 of a good triac show infinite resistance in either direction of ohmmeter polarity? _____
   _____
   _____

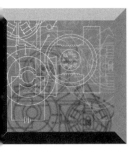

Activity 71–Universal Motor Speed Control

Name _____ Date _____ Score _____

Objectives

An important advantage of a universal (ac/dc) motor is the ease of speed control. Since this type of motor is similar to a series-wound dc motor, it has many of the same characteristics. The universal motor has a brush/commutator assembly, with the armature circuit connected in series with the field windings. By varying the voltage applied to universal motors, their speed can be varied from zero to maximum.

In the activity, you will observe the speed control characteristics of a universal motor. The circuit used for this purpose will be a gate-controlled triac. The triac is a semiconductor device whose conduction can be varied by a trigger voltage applied to its gate. A silicon controlled rectifier (SCR) could also be used as a speed control device for a universal motor. Speed control circuits like the one you will construct are used for many applications, such as electric drills, sewing machines, electric mixers, and several industrial applications.

A commercial light dimmer control can be substituted for the control circuit of this activity, if desired.

Equipment and Materials

Multimeter

- Universal motor

Resistors—4700 Ω, 47 kΩ, 68 kΩ

- Potentiometer—500 kΩ
- Capacitors (2)—0.1 μF

Diac—GE-X13

- Triac—GE-X12

Procedure

1. Construct the universal motor speed control circuit shown in Figure 71-1.

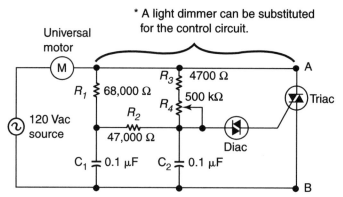

Figure 71-1. Speed control circuit for a universal motor.

2. Connect your multimeter, set to measure voltage, across the triac (point A to point B) to observe the voltage across the device.

3. Adjust the 500-kΩ speed control potentiometer through its range to check the circuit for operation. The motor speed should vary from zero to rated r/min.

4. Monitor the voltage across the triac as the potentiometer is adjusted. Observe the voltages across the triac and the motor at the voltage levels indicated in Figure 71-2. Record these voltages.

Voltage across motor	Voltage across triac
20 Vac	
40 Vac	
60 Vac	
80 Vac	
100 Vac	
120 Vac	

Figure 71-2. Voltage comparison of motor and triac circuit.

Activity 71—Universal Motor Speed Control

5. Monitor the voltage across the potentiometer. Record the maximum voltage as the potentiometer is varied, and observe its effect on the speed of the motor.

Maximum voltage = _____ volts ac.

6. Monitor the voltage across capacitor C_2. Record the minimum and maximum voltage across C_2 as the potentiometer is varied.

Minimum voltage = _____ volts ac.

Maximum voltage = _____ volts ac.

7. Turn off the power and disassemble your circuit. Return all materials.

Analysis

1. From the data of Figure 71-2, what is the relationship of the voltage across the motor and the voltage across the triac?

_____

2. At what voltage across the motor does it begin to rotate? _____

_____

3. How does the triac functionally control the speed of the universal motor? _____

_____

4. From the data of Step 5, as potentiometer resistance increases, what happens to motor speed?

_____

5. From the data of Step 6, as the voltage across capacitor C_2 increases, what happens to motor speed? _____

_____

6. Why does this type of control allow the universal motor to produce more torque through its entire operational range than a rheostat control that reduces the applied ac voltage? _____

_____

Activity 72–Inverting Op-Amps

Name _____ Date _____ Score _____

Objectives

Applications of the op-amp are very common in control circuits. Integrated circuits of this type are used primarily as high-gain amplifiers. The versatility of this amplifier makes it very useful in automatic control systems and instrumentation applications. With no external feedback circuit, an op-amp can achieve an open-loop gain that is around 100,000. With a feed-back circuit, gains can be achieved from 1 to 100,000 according to the selected resistance values.

In this activity, you will:

1. Construct a simple inverting op-amp, including both ac and dc signals.

2. Employ calculations that show voltage gain, current gain, and resistance value selection.

3. Design a simple operational amplifier that will achieve a specific amount of gain.

Equipment and Materials

- Multimeter
- Oscilloscope
- Split power supply (or two 9-V batteries)
- Variable low-voltage dc source
- AC signal generator
- Op-amp—μA741C
- Resistor—10 kΩ, 200 kΩ, 2 kΩ
- Capacitor—4-μF, 50 Vdc

Procedure

1. Construct the inverting dc op-amp of Figure 72-1. The power source of the μA741C is derived from a split power supply of two 9-V batteries, Figure 72-2, or variable dc supplies. Figure 72-3 shows the top view of the IC package.

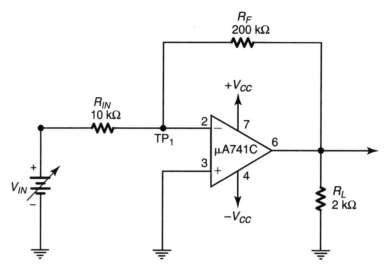

Figure 72-1. Inverting op-amp test circuit.

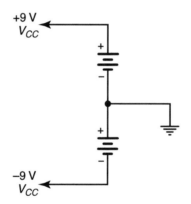

Figure 72-2. Split power supply.

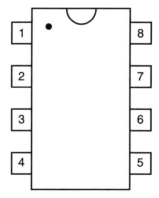

Figure 72-3. Pin layout of op-amp IC package.

2. Adjust the dc input source to produce 0.1 Vdc.

3. Energize the IC by turning on or connecting the split power supply. Measure and record the output voltage across R_L.

$V_{R_L} =$ _____ Vdc.

4. How does the polarity of the output compare with that of the input? _____

_____

5. Carefully increase the V_{IN} voltage to 0.3 V. Measure and record the output.

Output = _____ Vdc.

6. The dc voltage gain of an operational amplifier is determined by the formula $A_V = -R_F/R_{IN}$. The negative sign of R_F denotes the inversion function.

7. Calculate the gain factor for 0.3 V input.

Voltage gain = _____.

8. Test the measured value of Steps 3 and 5 with calculated values. How do they compare? How do you account for any differences between calculated and measured values? _____

_____

_____

9. Using the measured voltage across R_L in Step 5, calculate the load current passing through.

Calculated $I_L =$ _____ mA.

10. Measure the load current and record your findings.

Measured I_L _____ mA.

11. How do the measured and calculated values compare? _____

_____

12. Measure the input current at test point 1. Then calculate the input current with the formula $I_{IN} = V_{IN}/(R_S + R_{IN})$. We will assume that the resistance of a regulated power supply R_S is quite small and not applicable in this case. How does the measured value compare with the calculated value? _____

13. The current gain (A_I) of an inverting dc amplifier is determined by the formula $A_I = I_{OUT}/I_{IN}$. Calculate this value.

$A_I =$ _____.

14. Turn off the power supply and disconnect the dc input source V_{IN}. Connect a 4-µF capacitor in series with R_{IN} and an ac signal source of 400 Hz. Adjust the output of the signal source to its lowest value. Connect an oscilloscope across output resistor R_L.

15. Turn on the power supply and signal source. Carefully adjust the signal source input until it produces its maximum *undistorted* output. Record the peak-to-peak output voltage.

Output voltage = _____ V p-p.

16. Measure and record the input voltage at test point 1.

Input voltage = _____ V p-p.

17. Calculate the ac voltage gain of the amplifier using the formula $A_V = -R_F/R_{IN}$.

Voltage gain (A_V) = _____.

18. Turn off the power and disassemble your circuit. Return all materials.

Analysis

1. If the source resistance R_S were 500 Ω, how would it alter I_L in Step 10? _____

_____

2. If R_F were increased to 250 kΩ, how would it alter the voltage gain (A_V) in Step 7? _____

_____

3. If R_F were changed to 50 kΩ, how would it alter the voltage gain (A_V) in Step 7? _____

_____

4. How does the current gain (A_I) compare with voltage gain (A_V) in Steps 13 and 7? _____

_____

Activity 73–Noninverting Op-Amps

Name _____ Date _____ Score _____

Objectives

Circuits frequently need amplifiers that develop high voltage and current gains without signal inversion. The noninverting op-amp is specifically designed for this type of application. Both ac and dc signals can be amplified by this circuit, with the input and output remaining in phase. The input resistance or impedance of this amplifier is quiet high, with typical values exceeding 100 MΩ. With the input signal applied to the noninverting input terminal, voltage gain is dependent upon the values of R_{IN} and R_F. The output signal of this amplifier is developed across the load resistor, R_L. Typical R_L values are 35 to 50Ω.

In this activity, you will:

1. Connect a simple noninverting op-amp and observe its operation.

2. Calculate amplifier gain using both the resistance method and the voltage method.

3. Design a simple noninverting op-amp and alter its gain by changing the resistance values.

Equipment and Materials

Multimeter

Oscilloscope

Split power supply (or two 9-V batteries)

Variable dc source (or a 1.5-V C cell with a potentiometer)

AC signal generator

Op-amp—µA741C

Resistors—2 kΩ, 10 kΩ, 200 kΩ (2)

Capacitor—0.01 mF, 100 Vdc

Procedure

1. Construct the noninverting op-amp of Figure 73-1.

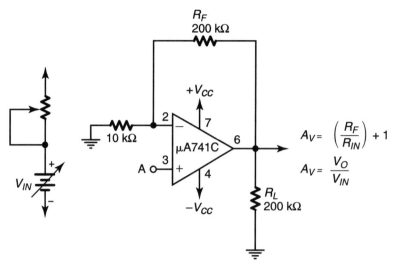

$$A_V = \left(\frac{R_F}{R_{IN}}\right) + 1$$

$$A_V = \frac{V_O}{V_{IN}}$$

Figure 73-1. DC noninverting op-amp test circuit.

2. Adjust the voltage input source to 0.1 V input and connect it to point A. Turn on the IC power source, and measure and record the output voltage developed across R_L.

$V_{OUT} = $ _____ Vdc.

3. Using the voltage amplification formula, calculate the A_V for the circuit using measured voltage values.

$A_V = $ _____.

4. Using the A_V formula for resistance, calculate the gain using the circuit values for R_{IN} and R_F.

$A_V = $ _____.

5. If there is a significant difference in values of these two voltage calculations, measure the actual value resistance of R_{IN} and R_F. Then calculate A_V using these actual values.

6. Connect two 200-kΩ resistors in series to make a 400-kΩ R_F. Repeat 2, 3, and 4 for the new value of R_F. Record the measured value and calculated values.

$V_{OUT} = $ _____ V.

A_V (by voltage) = _____.

A_V (by resistance) = _____.

7. Connect the two 200-kΩ resistors in parallel and repeat Step 6 for V_{IN} of 0.3 V. Record the measured and calculated values.

$V_{OUT} = $ _____ V.

A_V (by voltage) = _____.

A_V (by resistance) = _____.

8. Alter the dc amplifier of Figure 73-1 to conform with the ac amplifier of Figure 73-2. Resistors R_{IN} and R_1 should be of equal value.

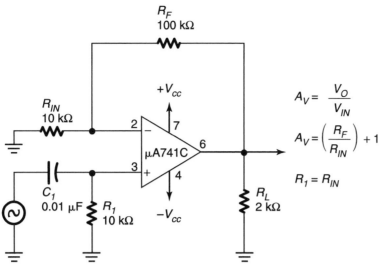

Figure 73-2. AC noninverting op-amp test circuit.

9. Turn on the ac signal generator and adjust it to produce a 400-Hz signal of 0.1 V p-p. Turn on the IC power supply.

10. Measure and record the output voltage appearing across R_L.

   $V_{OUT} = $ _____ V p-p.

11. Using the voltage formula calculate the voltage gain.

   $A_V = $ _____.

12. Calculate the voltage gain with the resistor formula.

   $A_V = $ _____.

13. Observe the phase relationship of the input signal while observing the output signal across R_L. At what input value does the output become distorted? _____

_____

14. Turn off the power and disassemble your circuit. Return all materials.

Analysis

1. What is the primary difference between an ac and a dc noninverting op-amp? _____

_____

2. Why is a capacitor used on the input of the ac amplifier? _____

_____

3. How would the value of the capacitor limit the amplifying frequency of the amplifier? _____

_____

4. Why does R_F influence voltage gain? _____

_____

Activity 74–Proximity Detectors

Name _____ Date _____ Score _____

Objectives

Proximity detectors are used to trigger an alarm or turn on a circuit by detecting physical change within a designated area. These circuits are designed to respond in some way to a change in capacitance. This change usually alters the frequency of an oscillator so that the output can be amplified and used to trigger a load control device. Typically, the circuit contains an oscillator, a tuned circuit, an amplifier, and a trigger device.

One of the most popular proximity detectors in operation today utilizes the loaded oscillator principle. In this type of circuit, the LC components of the oscillator are shunted to ground by an external capacitance. This capacitance is usually called the *sensor plate*. Moving a hand or producing a physical change near the sensing plate upsets the electrostatic field of the capacitance. This in turn alters the frequency of the oscillator or may load down the circuit in such a way as to quench the oscillation process. Liquid level detection, automatic thickness gauge testers, intruder alarm systems, counters, and switching circuits employ this type of detector.

In this activity, you will construct a loaded oscillator proximity detector. The oscillator is a Colpitts type with split capacitors in the collector circuit. The output of the oscillator is rectified and applied to a voltage comparator circuit where it is amplified by an op-amp. The amplified output is used to trigger a load control device. The 5 kΩ resistor in the comparator circuit is used to establish a balanced condition for detection. When the oscillator senses a change, it loads down and stops oscillating. This creates an imbalance, which is then detected by the comparator circuit.

Through this activity, you will test the operation of a loaded oscillator circuit and observe typical waveforms. The test procedure used to analyze circuit operation is also used to troubleshoot the circuit. Through this testing experience, you will become familiar with the oscillator loading principle and be able to follow a logical troubleshooting procedure.

Equipment and Materials

Multimeter

Oscilloscope

DC power supply—0-10 V, 1 A

Op-amp—LM3900

Transistor—2N3397

Coil—10 mH

Diode—1N4004

Light-emitting diode

Resistors—6.8 kΩ, 10 kΩ (2), 68 kΩ, 1 MΩ (2), 1/4 W

Resistor—470 Ω, 1/8 W

Potentiometers—5 kΩ, 500 kΩ, 2 W

Capacitor—0.1 μF, 200 Vdc

Capacitors—47 pF, 100 pF, 0.01 μF, 100 Vdc

SPST toggle switches (2)

Sensor plate (6″ × 6″ metal)

Optional according to the selected load control circuit

- Transistor—2N3397
- Relay—12V, 125 Ω (Guardian 1335-2C-120D or equivalent)
- No. 47 lamp with socket
- Resistors—47 Ω, 220 Ω, 1/2 W
- SCR (General Electric C-122D or equivalent)
- AC source—6.3 V, 60 Hz

Procedure

1. Construct the proximity detector of Figure 74-1.

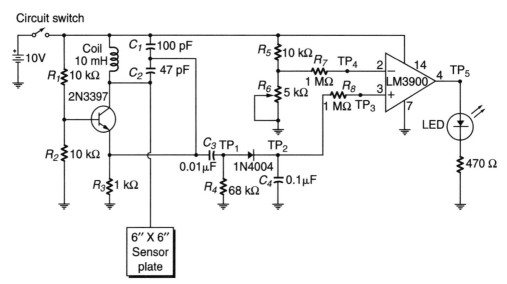

Figure 74-1. Proximity detector.

2. Close the circuit switch and adjust the 5 kΩ potentiometer for illumination of the LED. There should be a specific point where the LED turns on. Increasing resistance will cause it to remain on, while decreasing the resistance causes it to turn off. The point just before turn-on represents the most sensitive detection setting of the potentiometer. Adjust the potentiometer several times to locate the most sensitive point.

3. If the circuit is not operating properly, perform Steps 4, 8, 9, 10, and 11 first to determine which part of the circuit is malfunctioning. If the circuit is operating satisfactorily, follow the procedure steps in order and record the indicated data.

4. Prepare an oscilloscope for operation and connect it to test point 1 (TP$_1$). Make a sketch of the observed waveform, Figure 74-2.

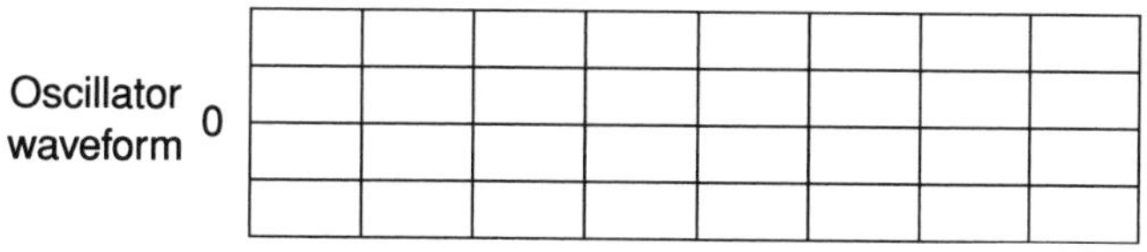

Oscilloscope 0
waveform

Figure 74-2. Observed waveform at test point.

5. While observing the oscilloscope, touch the sensor plate. What influence does this have on the waveform? _____

_____

6. Move your hand over the sensor plate while observing the oscilloscope. What influence does this have on the waveform? _____

_____

7. Determine the frequency of the oscillator with your oscilloscope.

   Oscillator frequency _____ Hz.

8. Measure and record the voltages at the emitter, base, and collector of the oscillator transistor Q_1. Compare your measurements with those indicated on the schematic diagram.

   Emitter voltage = _____ V.

   Base voltage = _____ V.

   Collector voltage = _____ V.

9. Measure and record the dc output voltage at test point 2 (TP$_2$).

   Unloaded oscillator output voltage = _____ V.

10. Touch the sensor plate and record the measured voltage at test point 2. This represents the loaded output voltage of the oscillator.

    Loaded output voltage = _____ V.

    The output of the oscillator will change approximately 2 V between the loaded and unloaded conditions of operation.

11. Measure and record input voltages at TP$_3$ and TP$_4$ when the LED is on. This indicates a balanced condition.

    TP$_3$ = _____ V.

    TP$_4$ = _____ V.

12. Measure and record the voltages at which the circuit is unbalanced.

    TP$_3$ = _____ V.

    TP$_4$ = _____ V.

13. Measure and record the output voltage of the circuit at test point 5 (TP$_5$) for the balanced and unbalanced conditions.

    The total change in voltage is _____ V.

14. Select one of the alternate load control circuits in Figure 74-3 and construct it. Connect it to the output of the proximity circuit.

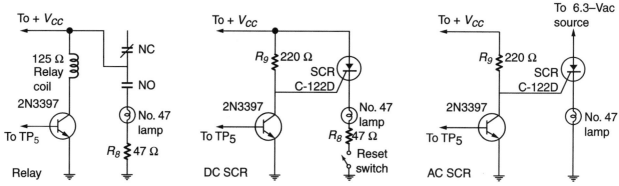

Figure 74-3. Alternate load control circuit.

15. Close the circuit switch, adjust the balance potentiometer, and test the control capacity of the completed load control circuit.

16. Open the circuit switch. Disconnect the circuit and return all parts.

Analysis

1. Describe what is meant by loaded and unloaded oscillator conditions. _____

    _____

2. What is the function of the IC op-amp in the proximity circuit of Figure 74-1? _____

    _____

3. Draw a block diagram representing the major parts of the proximity circuit constructed in Step 15.

Activity 75–Reed Switches

Name _____ Date _____ Score _____

Objectives

Reed switches are devices that respond to a controlled magnetic field. They are designed to close or open when exposed to either a permanent magnetic field or to an electromagnetic field.

The contacts of a reed switch are housed inside a hermetically sealed glass tube. When actuated, contact sparks are isolated from the outside environment. Industrial applications of this device are quite numerous in explosive area and dirt-prone environments. They are used to verify *z* axis movement on some robots.

A reed switch contains two flat metal strips, or reeds, housed in a hollow glass tube filled with an inert gas. When the reeds are exposed to a magnetic field, they are forced together, making or breaking contact depending on their design. The normally closed switch breaks contact when exposed to a magnetic field. Normally open contacts, by comparison, are forced closed when exposed to a magnetic field.

In this activity, you will construct a simple electrical circuit and control its action by changes in magnetic field strength. Through this experience you will gain some insight into the operation, sensitivity, and control action of a reed switch.

Equipment and Materials

- Multimeter
- DC power supply—0.5 V, 1.0 A
- AC power source—6.3 V, 60 Hz.
- Reed switch (GE-X7 or equivalent)
- Bobbin-wound, reed switch coil
- Reed switch magnet
- Lamp—No. 47
- Switch—SPST
- Circuit construction board

Procedure

1. Construct the reed-switch circuit of Figure 75-1.

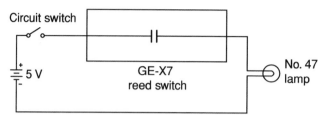

Figure 75-1. Reed switch test circuit for observing permanent magnet actuation.

2. Turn on the circuit switch and move the small permanent magnet near the reed switch. Try the magnet at a right angle with reeds, parallel with the reeds, etc. Which type of magnet orientation produces the best control capability?

3. Place the reed switch in the center of the bobbin-wound coil. Connect the circuit of Figure 75-2 to test the electromagnetic action of the reed switch.

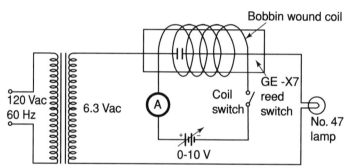

Figure 75-2. Reed switch test circuit for observing electromagnetic actuation.

4. Turn on the ac power source and the dc coil-control circuit.

5. Starting at zero volts dc, gradually increase the dc source voltage until the reed switch is actuated. The meter should be set at a high current range until an approximate actuating value is determined. Run at least two trial tests to determine the actuating current needed to energize the switch.

   DC actuating current = _____ mA.

6. In some applications, the actuating coil is used to increase the sensitivity of the reed switch by producing a partial field. Increase the dc coil current close to the actuating value. Then place the permanent magnet near the coil to actuate the reed. In this case, the switch can have some degree of variable sensitivity.

7. Turn off the ac and dc power sources for the circuit of Figure 75-2. Use 5 volts dc to supply the coil. The primary side of the coil should be connected to a variable ac transformer.

8. Turn on both the ac and dc sources, then increase the ac voltage applied to the coil. Measure the applied ac voltage. Do not increase the voltage to a value that will cause the reed coil to overheat.

   How does the reed switch respond to ac compared with dc? _____

   _____

9. Turn off the ac and dc power sources and disconnect the circuit. Return all components.

Analysis

1. How could a reed switch be used to control a pump motor in a liquid sump tank? _____

_____

_____

2. Make a sketch of this circuit.

3. How could a reed switch be used to control a circuit in an explosive area without danger of an explosion? _____

_____

_____

Activity 76–Touch Switch Circuit

Name _____ Date _____ Score _____

Objectives

A number of switches used to control industrial machinery do not employ any moving mechanical parts. This classification of switch is commonly called a touch switch. It achieves control by someone simply touching a seemingly solid button or plastic cover plate. The actual circuit is a capacitance switch that responds to a physical change in circuit capacitance caused by touching a finger to a certain area. In many cases the circuit can be adjusted so that switching is achieved by simply placing a finger in close proximity to the actual contact area. Switches of this type are very reliable and in many cases, used for safety reasons in explosionproof installations.

In this activity, you will construct a FET/op-amp touch switch that triggers an SCR. Load control of a lamp is achieved without moving parts. When the source of the load control circuit employs dc, manual resetting is required. Alarm circuits frequently employ this type of control. When the source of the SCR is ac, the touch switch circuit has automatic reset after each half cycle of operation.

Through this activity you will test the signal path from the FET through the op-amp and ultimately to the SCR. This signal tracing process is often used to troubleshoot circuit of a similar nature. The touch switch employs a number of components that you have studied in other applications. The combination of FET, op-amp, and SCR does, however, represent a rather unusual circuit application.

Equipment and Materials

- Multimeter

- Split power supply—0–5 V, 0–5 V, 1 A (or two 9-V batteries)

- AC source 6.3-V, 60-Hz

- SCR (General Electric C-122D or equivalent)

- Op-amp—µA741C

- FET (General Electric FET-1 or equivalent)

- Potentiometer—5 kΩ, 2 W

- Resistors—120 Ω, 4.7 kΩ, 10 kΩ, 56 kΩ, 220 kΩ, 10 MΩ, 1/2 W

- No. 47 lamp with socket

- SPST toggle switch

- IC circuit construction board

- Piece of metal of any gauge (1″ square)

Procedure

1. Construct the touch switch of Figure 76-1.

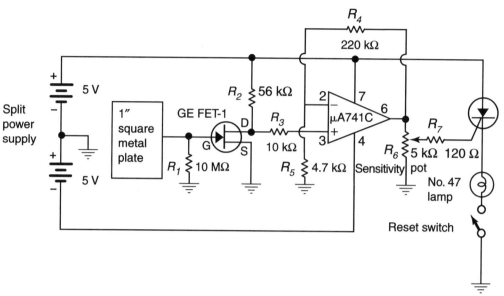

Figure 76-1. Touch switch circuit.

2. Note that a split 5-V power supply is used to supply power to the μA741C and other parts of the circuit. Two 9-V batteries can be used in place of the variable dc power supplies.

3. Close the reset switch and touch the 1" square metal plate. Adjust the 5 kΩ sensitivity control while touching the plate. The lamp should turn on at a certain setting of the potentiometer.

4. Remove your finger from the touch plate and open the reset switch momentarily, then turn on the switch. If the sensitivity is properly set, the lamp should remain off until the metal plate is touched. You may need to try several attempts at setting the sensitivity to its best range. The circuit must be reset each time.

5. If the circuit does not work properly you may want to make the following test as a troubleshooting procedure. If the circuit is working properly the following tests are made to analyze the operation of the circuit.

6. With power applied to the circuit, connect a multimeter between the drain and ground of the FET. Touch the metal plate while observing this voltage. As a general rule the voltage change is very small.

   Change in drain voltage is _____ to _____ V.

7. The change in drain voltage is then applied to the positive or noninverting input of the μA741C. This voltage is amplified and appears as the output voltage at pin 6. Measure and record the IC output voltage change.

   Output voltage ranges from _____ to _____ V.

8. Output voltage from the op-amp is applied to the trigger sensitivity control of the SCR. This voltage controls the gate current needed to trigger the SCR into conduction. Measure and record the change in gate voltage.

   The range of gate voltage is _____ to _____ V.

9. Measure the voltage across the gate resistor when the SCR is being triggered.

   Gate triggering voltage is _____ V.

Activity 76—Touch Switch Circuit

10. Calculate the gate current needed to trigger the SCR.

$I_G =$ _____ mA.

11. Carefully disconnect the +5 volt dc line from the anode of the SCR. In its place connect a 6.3 volt ac source to between anode and ground. Turn on the reset switch.

12. Using the procedure outlined in Steps 3 and 4 adjust the sensitivity so that the SCR triggers when the metal plate is touched. How does the operation of the circuit differ from that of the dc SCR circuit? _____

_____

13. Turn off the circuit power supplies. Disconnect the circuit and return all components.

Analysis

1. Why does the SCR need to be reset after it has been triggered? _____

_____

2. What does touching the metal plate actually change in the FET? _____

_____

3. Why does an ac supply to the SCR have automatic resetting? _____

_____

Activity 77–Temperature Sensors

Name _____ Date _____ Score _____

Objectives

Temperature sensing circuits are widely used in industry as system control elements. Heat sensors, which are by far the most popular of all sensors, are commonly found in alarm circuits that detect changes in temperature at remote locations. Machinery, electronics equipment, and measuring instruments also employ temperature sensors to detect unusual operating conditions. The thermistor is a very popular sensor element for temperature detection equipment.

Thermistors are primarily classified as temperature sensitive resistors with an operating range of from –382°F (–230°C) to 1202°F (650°C). When used in bridge circuits with high-gain amplification, thermistors can detect temperature changes as small as 0.001°.

The physical makeup of a thermistor includes a mixture of nickel, manganese, and cobalt oxides formed into a piece of semiconductor material. These oxides are mixed together and fired to form a coherent nonporous material. The mixture is then formed into a variety of shapes. In this activity a tiny piece of the ceramic material is formed into a bead and enclosed in glass. Changes in temperature cause the resistance of the thermistor to produce a wide range of values. Typically, thermistors have a negative temperature coefficient. This means that an increase in temperature causes a decrease in resistance. Metal, by comparison, has a positive temperature coefficient. This means that a rise in temperature causes a corresponding increasing resistance.

In this activity, you will build a temperature detecting bridge circuit. The output of the bridge is then fed into an op-amp for high-gain amplification and increased sensitivity. Any imbalance in the bridge is detected by the amplifier and is used to trigger an output circuit. Circuits of this type are typically found in alarm circuits that function as heat detectors.

Equipment and Materials

- Multimeter
- Split dc power supply—±9 V, 1 A (or two 9-volt batteries)
- Resistors—47 Ω, 1 kΩ, 2.7 kΩ, 10 kΩ (2), 200 kΩ, 1/4 W
- Potentiometer—5 kΩ, 2 W
- Op-amp—μA741C
- Thermistor (Fenwal GB 32J2 or equivalent)
- SCR (GE C-122D or equivalent)
- No. 47 lamp with socket
- IC circuit construction board
- SPST toggle switch
- Decade resistance box (optional)

Procedure

1. Construct the thermistor bridge circuit of Figure 77-1. Note that a voltmeter is used to detect the amplified output of the circuit.

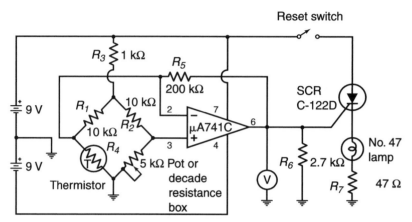

Figure 77-1. Thermistor bridge circuit.

2. Before connecting the thermistor into the circuit, measure its resistance. Try to avoid touching the glass part of the thermistor when attaching the test leads. With the meter connected, the resistance should stabilize after a few seconds.

   The stabilized resistance is _____ Ω.

3. While observing the meter, grasp the glass bead between your index finger and thumb. How does body temperature influence the resistance of the thermistor? _____
_____

4. Connect the thermistor to the bridge circuit as indicated. The reset switch should be in the off position and the meter should be adjusted to the zero center position. It should be in the 50 V or equivalent range.

5. Turn on the split power supplies and energize the circuit. Balance the bridge by adjusting the potentiometer to produce a zero indication of the meter. A decade resistance box can be used in place of the potentiometer.

6. After the bridge has been nearly balanced, you can switch the meter to a lower voltage range to improve the balancing accuracy.

7. To test the sensitivity of the circuit, place your finger near the glass bead of the thermistor while observing the meter. If the circuit is working properly, the meter should deflect upscale or in the positive direction. Within a few seconds it should return to the balanced indication when the thermistor has reached its stabilized resistance. If you touch the glass bead with your finger, it usually takes longer for it to return to the balanced state.

8. Try blowing on the thermistor or placing it near some heat producing source. Avoid temperatures over 302°F (150°C).

9. Close the reset switch of the SCR. Place your finger near the thermistor. While observing the meter, see how much voltage is needed to trigger the SCR into conduction. You must wait a few seconds for the thermistor to stabilize before resetting the SCR.

10. Try several sources of heat to turn on the second circuit.

11. Turn off the power supply and disconnect the circuit. Return all components.

Analysis

1. What type of temperature coefficient does the thermistor used in this activity have? _____

   _____

   _____

2. Explain how a change in thermistor resistance causes the bridge circuit to be imbalanced. ____

   _____

   _____

3. Why is it advantageous to use an op-amp with the bridge circuit of this activity? _____

   _____

   _____

Activity 78–Thermocouple Applications

Name _____ Date _____ Score _____

Objectives

Thermocouples are frequently used to measure temperatures in an industrial setting. Due to the relatively low voltage output associated with most thermocouples, amplification circuits are used to increase this output as well as to increase sensitivity. The resulting output of the amplification circuit is used to drive or activate readout device.

In this activity, you will see the thermocouple used to control the conductivity of a FET. The conductivity of the FET, in turn, controls the action of a single stage transistor amplifier and thus the current through a multimeter that is used as the readout device.

Equipment and Materials

- Digital multimeter
- Variable dc power supply—0-5 volts
- DC power supply—10 volts
- AC power supply—120 V
- Thermocouple—type J
- Resistors—5.6 Ω, 560 Ω
- Field-effect transistor—GE-FET-1
- NPN transistor—2N2405
- Connecting wires
- Heat cone—660 W

Procedure

1. Construct the circuit shown in Figure 78-1.

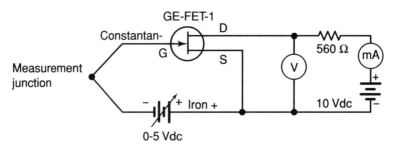

Figure 78-1. Type J thermocouple FET circuit.

2. Allow the measurement junction of the thermocouple to remain at room temperature. Alter the gate voltage of the FET to equal those listed in Figure 78-2. Record the source–drain current and voltage for each FET gate voltage value.

Gate voltage (V)	Source-drain current (mA)	Source-drain voltage (V)
0.2		
0.4		
0.6		
0.8		
1.0		
1.5		
2.0		

3. Adjust the gate voltage to 0.2 V.

4. Grasp the measurement junction of the thermocouple between your thumb and forefinger. Describe how this action affects the source–drain current and voltage of the FET as compared to the data gathered in Step 2 when the gate voltage was 0.2 V. _____

_____

5. Connect the 660 W heat cone to 120 Vac and allow it to warm up for about three minutes.

6. Position the measurement junction of the thermocouple inside the heat cone for a period of three minutes and record the source–drain current and voltage of the FET.

$I_{SD} =$ _____; V_{SD} _____.

7. How do the values of source–drain current and voltage recorded in Step 6 compare with the data gathered in Step 2 when the FET gate voltage was 0.2 Vdc? _____

_____

8. Disconnect the heat cone from the 120 Vac power supply. What effect does the removal of the heat cone have on the source–drain current and voltage? _____

_____

9. Construct the circuit illustrated in Figure 78-3.

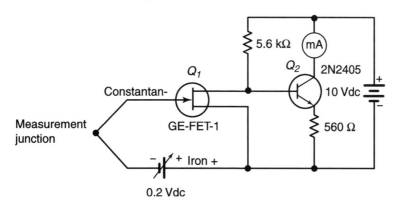

10. Record the collector current of Q_2 as displayed by the digital multimeter with the measurement junction of the thermocouple at room temperature.

    $I_C =$ _____.

11. Grasp the measurement junction of the thermocouple between your thumb and forefinger. Describe how this action affects the collector current of Q_2 as displayed on the digital multimeter. _____

_____

12. Place the measurement junction of the thermocouple inside the cone of the cool 660 W heater. Connect the 120 Vac to the heat cone and allow it to warm up for five minutes.

13. Record the collector current of Q_2 as displayed on the digital multimeter after the heat cone has warmed up.

    $I_C =$ _____.

14. How does the current recorded in Step 10 compare with the current recorded in Step 13? ____

_____

15. How do you explain the difference? _____

_____

16. If the digital multimeter used in the circuit shown in Step 9 was calibrated in degrees Fahrenheit, how could the circuit be used to measure temperature? _____

_____

17. Disassemble your circuit and return all components.

Analysis

1. Why is it sometimes necessary to use an amplifier when a thermocouple is used to measure temperature? _____

   _____

   _____

2. What was the readout device used in the circuit constructed in Step 9? _____

3. What would determine the maximum temperature that could be measured by the circuit in Step 9? _____

   _____

   _____

4. How did the action of the circuit used in Step 9 differ from that used in Step 2? _____

   _____

   _____

5. Explain how the circuit in Step 2 could be used to measure temperature. _____

   _____

   _____

Activity 79–Linear Variable Differential Transformer (LVDT) Detector

Name _____ Date _____ Score _____

Objectives

Frequently, it is not enough to measure only the amount of the displacement of an object. Often the direction of displacement, as well as the amount of displacement, must be indicated. When this becomes the case, a detector circuit similar to that shown in Figure 79-1 can be used in conjunction with the output of the linear variable differential transformer (LVDT).

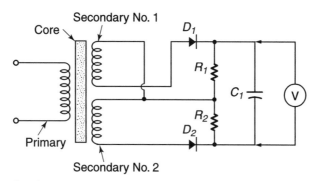

Figure 79-1. LVDT detector circuit.

In this detector circuit, the diodes conduct equally when the movable core is in the null position. This results in an equal and positive voltage drop across R_1 and R_2, causing the readout device (voltmeter) to indicate zero displacement. When displacement occurs, one diode will conduct more than the other, causing an upset in the balanced voltage across R_1 and R_2. The result is an output voltage, as measured by the voltmeter, whose polarity indicates the direction of displacement and whose magnitude indicates the amount of displacement. In this activity, you will examine the action of an LVDT detector circuit.

Equipment and Materials

Digital multimeter

Power supply—6 Vac

Air-core coils—100 turns, A.S. No. 16 wire (2)

Air-core coil—200 turns, A.S. No. 24 wire

Steel core—6" (15 cm) in length, 3/4" (1.91 cm) in diameter

Diodes—1N4004 (2)

Resistors—1 kΩ (2)

Capacitor—1 μF, 25 Vdc

Connecting wires

SPST switch

315

Procedure

1. Construct the circuit illustrated in Figure 79-2.

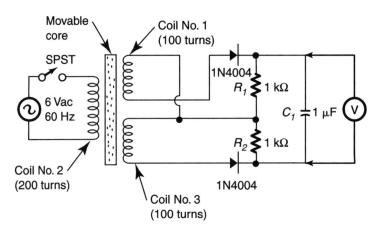

Figure 79-2. LVDT detector test circuit.

2. Close the SPST switch and slide the movable core until the meter indicates zero volts.

3. Moving the core in the direction and by the amount indicated in Figure 79-3, complete the tale by recording the voltage output of the detector circuit along with the proper voltage polarity.

Direction of movement from null position	Amount of movement Inches	Centimeters	Voltage output	Polarity (+ or −)
Left	1/16	0.159		
Left	1/8	0.318		
Left	3/16	0.477		
Left	1/4	0.635		
Left	5/16	0.793		
Left	3/8	0.953		
Return to null position				
Right	1/16	0.159		
Right	1/8	0.318		
Right	3/16	0.477		
Right	1/4	0.635		
Right	5/16	0.793		
Right	3/8	0.953		

Figure 79-3. Output polarity and magnitude of LVDT test circuit.

Activity 79—Linear Variable Differential Transformer (LVDT) Detector

4. How do the voltages generated by moving the core to the left compare with the voltages generated when the core is moved to the right? _____

_____

5. Ideally, what is the minimum displacement of the movable core that would result in a voltage output? _____

_____

6. Disassemble your circuit and return all components.

Analysis

1. What is meant by coils being connected in series to oppose? _____

_____

2. Explain how the detector circuit used in conjunction with the LVDT enables the direction of displacement to be measured. _____

_____

_____

3. When is it necessary to measure direction, as well as amount of displacement? _____

_____

_____

4. How would the null position of the core be affected if R_1 of the detector circuit were a value different from R_2? _____

_____

_____

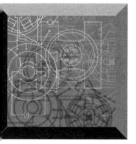

Activity 80–Strain Gauges

Name _____ Date _____ Score _____

Objectives

The strain gauge is a commonly used resistive transducer. It is used in the measurement of displacement, torque, weight, pressure, and other force parameters. A strain gauge has a fine-wire element looped back and forth on a mounting plate, which is usually cemented to the member undergoing stress. The extra length due to the looping increases the effect of any stress applied lengthwise. A tensile stress stretches the wire, increasing its length and, therefore, its resistance.

The measure of the sensitivity of a material to strain is called the gauge factor (GF), and is the ratio of the change in resistance (R/R) to the change in length (L/L) or:

$$GF = \frac{\Delta R/R}{\Delta L/L}$$

The initial resistance (R) is typically around 120 Ω, and the gauge factor may run from -12 to $+6$. A gauge factor of 2 is common for most wire strain gauges.

The more recent semiconductor strain gauges have greater sensitivity and can be used exactly like a metallic strain gauge. Semiconductor strain gauges have a gauge factor roughly 50 times greater (around 100) than the metallic types. This large gauge factor is accompanied by a thermal rate of change of resistance also approximately 50 times higher than in older gauges. Thus, the semiconductor strain gauge is as stable as the metallic type, but has a much higher output. Simpler temperature compensation methods can be applied to the semiconductor strain gauge, making it more useful and extending its range to the measurement of small values of microstrain (microinches/inch of length), which could not be measured with earlier gauges.

In this activity, a strain gauge used in a bridge network will be observed. This method can be used for most strain gauge measurements.

Equipment and Materials

Multimeter

Galvanometer

Resistor—1 kΩ, 100 kΩ

Strain gauge

Variable resistor (calibrated)—20 kΩ

DC power source

Procedure

1. Construct the strain gauge balance detector circuit shown in Figure 80-1.

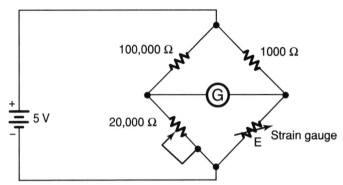

Figure 80-1. Strain gauge balance detector test circuit.

2. Measure the resistance of the strain gauge with the meter.

    Resistance = _____ Ω.

3. While your lab partner bends the metal mounting plate, monitor the null detector and observe the effect on the meter reading.

4. Disassemble your circuit and return all components.

Analysis

1. Write an analysis of the operation of the strain gauge. Include in your analysis your observation of the bridge network as a measuring method for the strain gauge. _____

    _____
    _____
    _____
    _____
    _____

Activity 81–Thermistors

Name _____ Date _____ Score _____

Objectives

Thermistors are commonly used as temperature sensitive transducers. Their most important property is that their resistance decreases with increases in temperature. In this activity, you will observe the characteristics of one type of thermistor.

Equipment and Materials

Multimeter

• Variable dc power source

• Thermistor—JA41J1

Resistor—1000 Ω

• Lamp—60 W

Procedure

1. Construct the circuit shown in Figure 81-1.

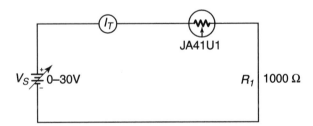

Figure 81-1. Thermistor characteristics test circuit.

2. Complete the Figure 81-2 relative to the circuit of Figure 81-1 with the thermistor at room temperature: (Note: Allow about two minutes between source voltage changes.)

Source Voltage (V_S)	I_T	Voltage Across Thermistor (V_T)	Calculated Resistance of Thermistor (V_T/I_T)	(V_{R_1})	Calculated Circuit Resistance (V_S/I_T)	Resistance Change of Thermistor
0.5						
1.0						
2.0						
3.0						
4.0						
5.0						
10						
15						
20						
25						
30						

Figure 81-2. Thermistor characteristics (room temperature).

3. Repeat Step 2 using a 6-watt lamp placed about one inch from the thermistor. This will increase the temperature and change the characteristics of the thermistor. (Again, allow about two minutes between voltage changes.) Record your data in Figure 81-3.

(V_S)	I_T	(V_T)	Calculated Resistance of Thermistor	(V_{R_1})	Calculated Circuit Resistance	Resistance Change of Thermistor
0.5						
1.0						
2.0						
3.0						
4.0						
5.0						
10						
15						
20						
25						
30						

Figure 81-3. Thermistor characteristics (heated).

4. Allow the thermistor to return to room temperature.

5. Apply 0.5 volts dc to the circuit and record the current caused by the circuit with the thermistor at room temperature.

Current = _____ mA.

6. Grasp the thermistor with your hand for three minutes and observe what happens to the current in the circuit.

7. Disassemble your circuit and return all components.

Activity 81—Thermistors

Analysis

1. Draw a current versus voltage graph to illustrate the characteristics of the thermistor from the data of Figures 81-2 and 81-3.

2. Briefly discuss the characteristics of a thermistor. _____

_____

_____

3. Compare the data of Figures 81-2 and 81-3 by noting whether each of the following measurements increased, decreased, or remained the same with an increase in temperature.

    a. Voltage across R_1 (V_{R_1}): _____

    b. Circuit current (I_T): _____

    c. Voltage across thermistor (V_T): _____

    d. Resistance of thermistor: _____

    e. Circuit resistance: _____

    f. Resistance changes of thermistor: _____

4. In Step 2, why is it necessary to allow two minutes between source voltage changes? _____

_____

_____

5. In Step 6, what happened to the circuit current when the thermistor was heated? _____

_____

_____

_____

Activity 82–Motor-Driven Timers

Name _____ Date _____ Score _____

Objectives

The motor-driven timer provides a wide variety of timing actions for industrial circuit applications. In its simplest form, this timer has an electric drive motor, ratchet release coil, and a ratchet dial that is held stationary until released. When the timing cycle reaches its set-time, the ratchet is released and the dial resets itself by spring action. Both on delay and off delay reset timers are available. More sophisticated reset timers permit a wide range of timing operations in a single unit.

In this activity, a simple rest timer will be used to build a load control circuit that produces either interval or delay timing operations. Initially, when the timer is energized by the control switch, load A is turned on and load B is turned off. After the expired time setting or the *time out* condition has been reached, load A is turned off and load B is turned on.

Equipment and Materials

- AC power supply—120 V
- Reset timer (Eagle Signal HD-50 Series)
- Incandescent lamps with sockets—7.5 W (2)
- SPST toggle switch

Safety

Be very careful when working with high voltages.

Procedure

1. Refer to the timer manufacturer's product manual before attempting to complete this activity.

2. Wire the reset timer for the *on delay* operation of Figure 82-1.

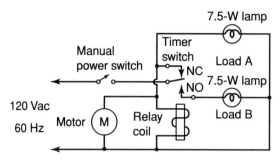

Figure 82-1. Reset timer with on delay operation.

3. Adjust the time setting dial by pulling it out and turning it to a desired setting. Releasing the dial automatically locks it in the new setting.

4. Turn on the manual power switch and describe the operating condition of loads A and B. ____

_____

_____

5. When the dial setting trips, describe the operating condition of the load and the motor. _____

_____

_____

6. Start a new condition cycle by opening the control switch momentarily, then turning it on again. After the cycle has been in operation for a few seconds, momentarily open the control switch, then close it. What action does this initiate? _____

_____

_____

7. Disassemble your circuit and return all components.

Analysis

1. What are some industrial applications of a reset timer? _____

_____

_____

2. Could this timer be modified to achieve a different function of some type? Explain. _____

_____

_____

3. Explain the difference between *interval* and *delay* timers. _____

_____

_____

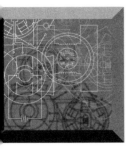

Activity 83–Motor-Driven Repeat Cycle Timers

Name _____ Date _____ Score _____

Objectives

Industrial applications of the motor-driven repeat cycle timers are very common. With this timing device, it is possible to control a number of specific operations in certain periods in a desired sequence. The operational time of this device is primarily based upon the rotational speed of a series of cams that actuate snap-action switches. The switching position of the cam can usually be altered within a range of from 2% to 5% of a single revolution of the shaft. The timer motor rotates the shaft through a gear train. Each cam actuated switch can be used to control a separate load device.

In this activity, you will connect up a simple repeat cycle timer and test its operational cycle. Sequence changes and cycle timing alterations will also be made. As a general rule, repeat cycle timer cam adjustments vary to some extent with each manufacturer. The timer used in this activity is only representative of one adjustment method. Specific adjustment procedures for timer devices must conform with the directions supplied by the manufacturer.

Equipment and Materials

AC power source—120 V

Motor-driven repeat cycle timer (Automatic Timing & Controls Co. Series 324C 02-B2D-R-1-A-01-X or an equivalent General Time Corp. RJ Series Unit)

Incandescent lamp and socket—7.5 W, 120 V, 60 Hz

SPST switch

Safety

Be sure to wear eye protection while electric motors are in operation. Also, be very careful when working with high voltages.

Procedure

1. Refer to the manufacturer's data sheet for the repeat cycle timer, Figure 83-1, before proceeding.

DF10 Series
- Solid state analog recycle timer circuitry
- 10A relay with DPDT contacts
- Fixed flash rate: Available from 10 to 240 FPM
- 12V to 120V input voltage range - Both AC and DC models
- UL File #E96739 (M)
- CSA File #LR62586-3

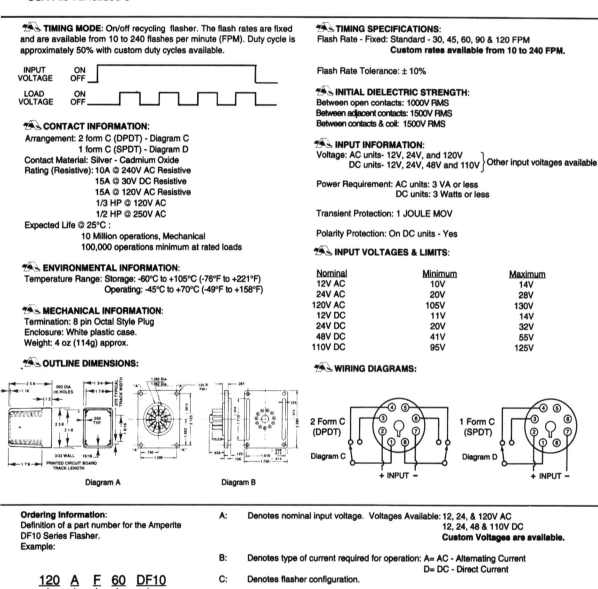

TIMING MODE: On/off recycling flasher. The flash rates are fixed and are available from 10 to 240 flashes per minute (FPM). Duty cycle is approximately 50% with custom duty cycles available.

CONTACT INFORMATION:
Arrangement: 2 form C (DPDT) - Diagram C
1 form C (SPDT) - Diagram D
Contact Material: Silver - Cadmium Oxide
Rating (Resistive): 10A @ 240V AC Resistive
15A @ 30V DC Resistive
15A @ 120V AC Resistive
1/3 HP @ 120V AC
1/2 HP @ 250V AC
Expected Life @ 25°C :
10 Million operations, Mechanical
100,000 operations minimum at rated loads

ENVIRONMENTAL INFORMATION:
Temperature Range: Storage: -60°C to +105°C (-76°F to +221°F)
Operating: -45°C to +70°C (-49°F to +158°F)

MECHANICAL INFORMATION:
Termination: 8 pin Octal Style Plug
Enclosure: White plastic case.
Weight: 4 oz (114g) approx.

OUTLINE DIMENSIONS:

Diagram A   Diagram B

TIMING SPECIFICATIONS:
Flash Rate - Fixed: Standard - 30, 45, 60, 90 & 120 FPM
Custom rates available from 10 to 240 FPM.

Flash Rate Tolerance: ± 10%

INITIAL DIELECTRIC STRENGTH:
Between open contacts: 1000V RMS
Between adjacent contacts: 1500V RMS
Between contacts & coil: 1500V RMS

INPUT INFORMATION:
Voltage: AC units- 12V, 24V, and 120V
DC units- 12V, 24V, 48V and 110V } Other input voltages available

Power Requirement: AC units: 3 VA or less
DC units: 3 Watts or less

Transient Protection: 1 JOULE MOV

Polarity Protection: On DC units - Yes

INPUT VOLTAGES & LIMITS:

Nominal	Minimum	Maximum
12V AC	10V	14V
24V AC	20V	28V
120V AC	105V	130V
12V DC	11V	14V
24V DC	20V	32V
48V DC	41V	55V
110V DC	95V	125V

WIRING DIAGRAMS:

2 Form C (DPDT)   1 Form C (SPDT)

Diagram C   Diagram D

+ INPUT −   + INPUT −

Ordering Information:
Definition of a part number for the Amperite DF10 Series Flasher.
Example:

120 A F 60 DF10
A   B C D   E

A: Denotes nominal input voltage. Voltages Available: 12, 24, & 120V AC
12, 24, 48 & 110V DC
Custom Voltages are available.

B: Denotes type of current required for operation: A= AC - Alternating Current
D= DC - Direct Current

C: Denotes flasher configuration.

D: Denotes flash rate. Standard rates are 30, 45, 60, 90 & 120 FPM.
Custom rates are also available from 10 to 240 FPM

E: Denotes 10A DPDT 2 form C flasher - DF10 Series.

Figure 83-1. Repeat cycle timer used for timing applications. Study this specification sheet. (Amperite Company, Inc.)

2. Construct the repeat cycle timer of Figure 83-2.

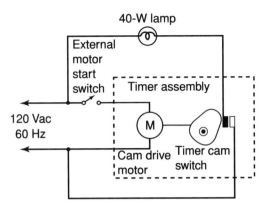

Figure 83-2. Repeat cycle timer.

3. Adjust the timer can for a 50% on and 50% off operational cycle. Then turn on the external motor start switch and test the operational cycle. The circuit should repeat itself as long as the external motor start switch is in the on position.

4. How long is the lamp on and off during a cycle of operation?
   On = _____ s; Off = _____ s.

5. Modify the repeat cycle timer to form a stop cycle operation as indicated in Figure 83-3.

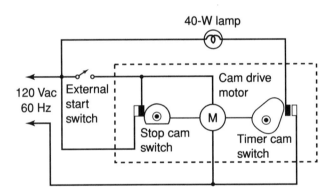

Figure 83-3. Timer with stop cycle operation.

6. To test the operation of this circuit, the external motor start switch must be turned on for at least 1% of the cycle time. It can then be opened. At the end of one operational cycle, the stop cam switch turns off the motor. To repeat the cycle, the external start switch must be turned on again.

7. If time permits, you might want to change the cam setting for a different range of control. If multiple cams are available on your timer, you can connect the unit to achieve control of alternate load.

8. Disassemble your circuit and return all components.

Analysis

1. Describe the basic action of the cam switch of a motor-driven cam timer. _____

   _____

   _____

2. How many control operations can the timer used in this experiment achieve during an operational cycle? _____

3. Describe an application for a repeat cycle timer. _____

   _____

   _____

Activity 84–Thermal Time-Delay Relays

Name _____ Date _____ Score _____

Objectives

Thermal relay devices are quite common in industrial equipment, and they have been used for industrial applications for a number of years and continue to be rather popular. Thermal timers are discrete devices designed to control circuit operation through the thermal expansion principle. When certain metals are heated, their expansion characteristic commonly causes a bending of the metal. Through this principle it is possible to force electrical contacts to come together or to open after a prolonged period. Devices that do this are called *thermal time-delay relays* because they are housed in a self-contained glass or metal package. Hermetically sealed units are not affected by atmospheric conditions, moisture, dust, or altitude.

The heat source of a typical thermal timer is provided by a resistive element mounted inside the enclosure. When the element is energized by an outside source, it causes normally open (NO) contacts to close and the normally closed (NC) contacts to become open. When the element source is removed, the contacts change back to their original state. As a general rule, the element can be energized by either ac or dc. Most timers of this type are designed to have a fixed timing action at the rated heater element voltage. At reduced voltage, however, timing action can be altered somewhat.

In this activity, you will construct an electrical circuit that utilizes a thermal relay. You will be able to test the actual expansion of a control element when it is energized. You will also alter the timing range of the device by decreasing the heating element energy source voltage.

Equipment and Materials

- Multimeter
- AC power source—6.3 V, 1 A, 60 Hz
- Variable dc source—0 to 6 V
- Thermal time-delay relay (Amperite 6NO30T or equivalent)
- 9-pin miniature tube socket
- Circuit construction board
- Incandescent lamp and socket—40 W, 120 V
- SPST toggle switch

Procedure

1. Study the thermal time-delay relay through the glass enclosure. See if you can identify the two heater pins. After deciding upon the two pins, measure the resistance with your multimeter.

   Coil resistance = _____ Ω.

2. Examine the specification sheets in Figure 84-1.

(Figure A)   (Figure B)   (Figure C)   (Figure D)

(Figure E)   (Figure F)   (Figure G)   (Figure H)

SOCKETS

TYPE	AMPERITE PART #	USED WITH THESE AMPERITE SERIES
Chassis Mount Octal Socket (Figure A)	Octal	C10, CI, DC10, DF10, G, GF
Surface Mount Octal Socket (Figure B)	8-Pin-SM	C10, CI, DC10, DF10, G, GF
Surface Mount 11-Pin Octal Style Socket (Figure C)	11-Pin-SM	CR10, CIR
Chassis Mount 9-Pin Socket (Figure D)	9-Pin-SM	G, GF
P.C. Mount 9-Pin Socket (Figure E)	9-Pin-PC	G, GF

QUICK CONNECTS

.110 Female Quick Connect Terminals (Figure F)	110 Female	B, BF, BR, C, CR, D
.250 Female Quick Connect Terminals (Figure G)	250 Female	B, BF, BR, C, CR, D, DF, DFW

BRACKETS

2-Screw Panel Mount Bracket (Figure H)	Panel Mount	B, BF, BR, C, CR, D, DF

A

Figure 84-1. Specification sheets. A—Sockets and mountings used with some types of thermal timers. B—Description of a *G Series* timer. (Amperite Company, Inc.)

Activity 84—Thermal Time-Delay Relays

G Series

- Hermetically sealed
- Delay on Make or Delay on Break timing modes
- Thermal device
- 3 AMP rating
- 1 - 115V input voltage range - works on AC or DC
- Isolated output contacts
- Fixed delay times only
- Initial and reset (release) delay device
- Long life
- UL File #E96739 (M)

TIMING MODE: Timing cycle begins upon application of power to the heater terminals. At the end of the initial delay time the relay contacts transfer and remain in a transfered state until input power is removed. When the heater input power is removed, the contacts transfer back to their original state at the end of a reset (release) delay period.

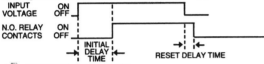

CONTACT INFORMATION:
Arrangement: 1 form A (SPST - Normally open) - Delay on Make
1 form B (SPST - Normally closed) - Delay on Break

Contact Material: Silver - Cadmium Oxide
Rating (Resistive): 3A @ 115V AC
Expected Life @ 25°C :
500,000 operations minimum at rated loads

ENVIRONMENTAL INFORMATION:
Temperature Range: Operating & storage: -55°C to +80°C, (-67°F to +176°F)

MECHANICAL INFORMATION:
Termination & Enclosure: Octal style, or 9-pin miniature style glass envelope. See Diagram A & B.
Weight: 1 oz (28g)

OUTLINE DIMENSIONS:

Diagram A
Standard Octal Base

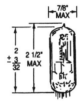

Diagram B
9-Pin Miniature Base

TIMING SPECIFICATIONS:
Timing - Fixed: 1 through 300 secs - (octal style) or
1 - 120 secs. (9-pin miniature style)
Timing Tolerance: ± 20% - **Tighter tolerances are available.**
Repeatability: ± 5%
Release Time: Contact factory
Timing Cycle Interrupt Transfer: none

INITIAL DIELECTRIC STRENGTH:

1 - 10 Second Type:	15 - 300 Second Type:
Between open contacts: 250V RMS	Between open contacts: 800V RMS
Between contacts & coil: 500V RMS	Between contacts & coil: 500V RMS

INPUT INFORMATION:
Voltage: AC or DC - 6V, 12V, 26V, 50V and 115V
(Other voltages are available)

Power Requirement: 2.0 Watts approx.

Transient Protection: impervious to transients

Polarity Protection: None required

INPUT VOLTAGES & LIMITS:

Nominal	Minimum	Maximum
6V AC/DC	4V	8V
12V AC/DC	10V	14V
26V AC/DC	22V	30V
50V AC/DC	42V	58V
115V AC/DC	90V	130V

WIRING DIAGRAMS:

Base Wiring 9-Pin Miniature
Pins 1 & 6 - Heater
Pins 3 & 4 - First Contact
Pins 8 & 9 - Second Contact

Bottom View
Diagram C

Base Wiring Standard Octal
Pins 2 & 3 - Heater
Pin 5 - First Contact
Pin 7 - Second Contact

Bottom View
Diagram D

Ordering Information:
Definition of a part number for the Amperite G Series Time Delay Relay.
Example:

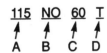

115 NO 60 T
↑   ↑   ↑   ↑
A   B   C   D

A: Denotes nominal input voltage. Voltages Available: 6, 12, 26, 50 & 115V AC/DC
Custom Voltages are available.

B: Denotes contact form: NO = normally open (Delay on Make) - 1 form A - SPST
C = normally closed (Delay on Break) - 1 form B - SPST

C: Denotes timing value: Factory preset time delays from 1 - 300 secs. are available (octal style) and 1 - 120 secs. (9-pin miniature style).

D: Denotes type of glass envelope: Blank = octal style. T = 9-pin miniature style.

B

Activity 84—Thermal Time-Delay Relays

3. Study the base diagram of the thermal time-delay device used in this activity. Make a sketch of the base diagram of this device in the space that follows.

4. If the thermal time-delay device used has a number, record it. _____. The number of this device often reveals a great deal about its operating characteristics. An Amperite type 24NO120T, for example, denotes a 24-V heater, normally open contacts, and a 120-second time delay.

5. Identify the operating characteristics of the thermal time-delay identified in Step 4.

6. In the space that follows, sketch a wiring diagram showing the thermal time-delay relay controlling a 7-W incandescent lamp. Energize the relay with 6.3 Vac and the lamp with 120 V at 60 Hz. Connect an SPST switch in series with the heater coil so that it can be easily turned on and off.

7. Have your instructor approve your circuit before applying power to the delay device or lamp.

8. Apply energy to the lamp part of the circuit first. Then turn on the heater coil switch and determine the delay time of the circuit.

   Delay time = _____ s.

9. Turn off the circuit and disconnect the 6.3-V, 60-Hz source from the heater coil. Connect a variable dc source in its place. Adjust the dc source to its zero volts position before connecting it to the heater coil part of the circuit.

10. Prepare the multimeter to measure dc voltage, and connect it to the variable dc source. Adjust the dc to 6 V. Then turn on the heater switch and record the delay time.

    Delay time = _____ s.

11. Prepare the multimeter to measure the dc current in the heater coil part of the circuit.

    Heater coil current = _____ A.

12. Using the applied heater voltage and current, calculate the operating coil resistance.

   Operating coil resistance = _____ Ω.

   How does this resistance compare with the cold coil resistance of Step 1? _____

   _____

   _____

13. Turn off the heater coil switch and adjust the source to 5 Vdc. Turn on the switch.

   Delay time = _____ s.

14. Measure and record the heater current when supplied with a 5-V source.

   Heater current = _____ A.

15. Plot a time delay versus heater coil voltage curve on the graph of Figure 84-2.

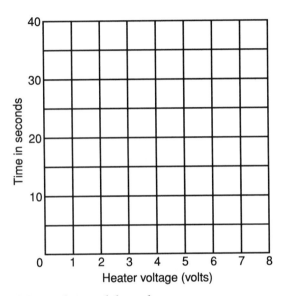

Figure 84-2. Characteristics of thermal time-delay relay.

16. Disassemble your circuit and return all components.

Analysis

1. What would be an application of the thermal time-delay device in an industrial installation?

   _____

   _____

2. What would be the advantages and disadvantages of using a thermal time-delay device in a delay circuit? _____

   _____

   _____

3. Since the dc heater coil voltage influences time delay, how could the delay time of this device be altered when a fixed dc source is used? _____

   _____

   _____

Activity 85–Time Constant Circuits

Name _____ Date _____ Score _____

Objectives

Inductance (L) opposes any change in current in a circuit, whereas capacitance (C) opposes any change in voltage. In both instances, the reaction time associated with inductance and capacitance in opposing changes to current and voltage is dependent upon resistance. The time, in seconds, required for a capacitor or an inductor to react to change is determined by a quantity known as a time constant. Circuits created with these time constants can be used to produce electronic timers.

In this activity, you will observe the time constant of a capacitor as it charges and discharges through a resistive path.

Equipment and Materials

Multimeter

Variable dc power supply

Resistors—220 kΩ, 100 kΩ

Capacitor—47 μF

SPST switches (2) (or one DPDT)

Connecting wires

Procedure

1. Construct the circuit in Figure 85-1.

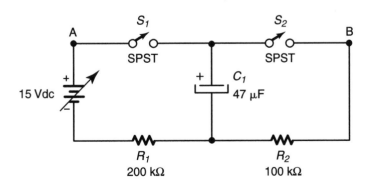

Figure 85-1. RC time constant test circuit.

2. When SPST switch S_1 is closed and SPST switch S_2 is open, the capacitor charges through R_1. After the capacitor is charged, it is discharged through R_2 by opening switch S_1 and closing switch S_2. Compute the time constant (τ) for charging and discharging the capacitor circuit.

   Charging $\tau =$ _____ s.

   Discharging $\tau =$ _____ s.

3. Compute the total time required for the capacitor to charge to 25 V when S_1 is closed and S_2 is open.

   Total charge time = _____ s.

4. Compute the total time required for the capacitor to discharge 25 V when S_1 is open and S_2 is closed.

   Total discharge time = _____ s.

5. Compute the total charge and discharge currents for the circuit shown in Figure 85-1.

   Total charge current = _____ mA

   Total discharge current = _____ mA

6. Prepare your multimeter to measure direct current in the 1-mA range and connect it in series with R_1 and C_1 at point A in the circuit shown in Figure 85-1. (Note: Both S_1 and S_2 must be open.)

7. Close S_1 and record the maximum charging current.

   Charging current = _____ mA.

8. Describe and explain the action of the circuit current as the capacitor charges. _____

   _____

9. How did the computed charging current in Step 5 compare with the measured charging current in Step 7? _____

10. Disconnect the meter, open S_1, and close S_2.

11. With the meter prepared to measure the direct current described in Step 6, connect this meter in series with C_1 and R_2 at point B in the circuit. (Note: Both S_1 and S_2 must be open.)

12. Close S_1 for five or six seconds. Open S_1, close S_2, and record the maximum discharge current.

   Discharge current = _____ mA.

13. Describe and explain the action of the current in the circuit as the capacitor discharged. _____

   _____

14. How did the computed discharge current in Step 5 compare with the measured discharge current in step 12? _____

   _____

15. How did the charge and discharge currents of the capacitor compare? Why were they so different? _____

   _____

16. Disconnect the meter and open S_2.

17. Prepare the meter to measure dc volts (25 V or higher) range and connect it across R_1 in the circuit.

18. Close S_1 and describe the voltage drop across R_1 as the capacitor charges. _____

_____

19. Disconnect the meter and place it across R_2.

20. Open S_1 and close S_2. Describe the voltage drop across R_2 as the capacitor discharges. _____

_____

21. As the capacitor is charging, the current and the voltage across R_1 decrease. When the charging current and voltage across R_1 reach zero, the capacitor is fully discharged. Likewise, as the capacitor discharges the discharge current and voltage drop across R_2 decrease. When the discharge current and voltage across R_2 reach zero, the capacitor is fully discharged.

22. Disconnect the meter and open S_1 and S_2.

23. Connect the meter across R_1, as described in Step 17.

24. Using a watch with a timer or a second hand, measure and record the time required for the capacitor to charge to 100% of the source voltage when S_1 is closed. This time corresponds to the time before the meter will measure zero voltage across R_1.

   Charging time = _____ s.

25. How did the measured time in Step 24 compare with the computed time in Step 3? _____

26. Connect the negative lead of the meter to the negative side of the capacitor. Do not connect the positive lead of the meter into the circuit at this time.

27. Open S_1 and quickly connect the positive lead of the meter to the positive side of the capacitor. Record the maximum voltage across the capacitor at the instant the positive lead is connected to the capacitor.

   Voltage across C_1 = _____ V

28. How did the voltage measured in Step 27 compare with the source voltage? _____

_____

Analysis

1. What is the time constant of an RL circuit? _____

2. What is the time constant of an RC circuit? _____

3. How many time constants are required for the current through an RL circuit to reach its maximum value? _____

4. How many time constants are required for the voltage across a capacitor in an RC circuit to equal the source voltage? _____

5. How can the length of an RL time constant be increased? _____
   Decreased? _____

6. How can the length of an RC time constant be increased? _____
   Decreased? _____

Activity 86–Digital Timers

Name _____ Date _____ Score _____

Objectives

Digital timers are a rather recent addition to the timer field. This type of timer employs a time base generator or oscillator, a counter, and a load driver. The counter part of the system frequently has several output terminals that permit some selection of time ranges. Control ranges of T, 2T, 4T, 8T, 16T, and 32T are typical. Each output is rated in terms of the time base (T) of the generator.

In this activity, you will employ the SE/NE555 timer IC as the time base generator of a simple digital timing system. The output of the generator will be counted and divided to some extent by the range of the system. Through this circuit investigation you will see and actually work with the basic parts of a simple digital timing system. Commercially designed timers of this type are usually housed in an enclosure that does not permit access to component parts. These timers generally do not employ moving parts and are all solid state. The timing period is very accurate and usually ranges from microseconds to hours.

Equipment and Materials

Power supply—5 Vdc

SE/NE555 IC

SN7490 IC

Resistors—390 Ω, 200 kΩ (2), 1/8 W

Capacitors—0.01 µF, 100 Vdc; 4 µF, 25 Vdc; 50 µF, 25 Vdc

Light-emitting diode

SPST toggle switches (or push button switches)

Circuit mounting board

Procedure

1. Construct the experimental digital timer circuit of Figure 86-1 by first connecting the time base generator. Do not energize the SN7490 at this time.

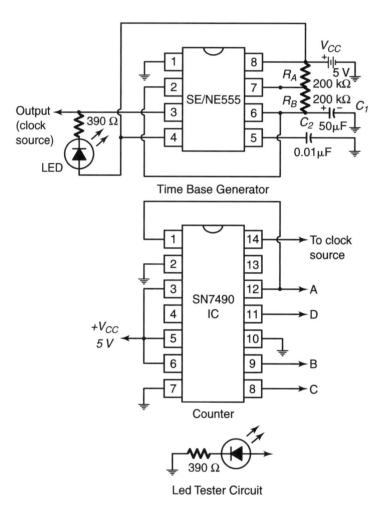

Figure 86-1.   Digital timer circuit.

2. Determine the charging time of the circuit by the $T_C = 0.693 \times (R_A + R_B) \times C_1$ formula. Test the accuracy of the circuit. Record your findings. _____

_____

3. Determine the discharge time of the circuit by the $T_D = 0.693 \times R_B \times C_1$ formula. Test the accuracy of the circuit. Record your findings._____

_____

4. Turn off the power supply and complete the SN7490 counter IC. Connect it to the same power source used by the 555. Connect the second LED indicator to the D output of the 7490 IC. Connect the IC clock source to the output (pin 3) of the 555.

5. Turn on the power source and count the number of timing cycles needed to energize the 7490 LED readout. What mathematical function does this represent?_____

_____

6. Turn off the power source and move the LED readout of the 7490 to output A. Turn on the power again and determine the cycles needed to energize the LED.

_____ timing cycles.

What mathematical function does this represent? _____

7. Turn off the power supply and alter the SN7490 circuitry as indicated in Figure 86-2. Change C_1 of the time base generator to 4 µF.

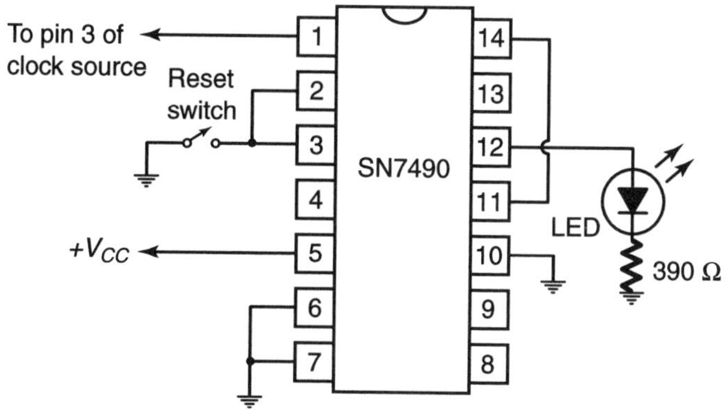

Figure 86-2.   Modified counter circuit.

8. Turn on the power supply and open the reset switch. Close the switch during a time when the time base generator LED is off. Then count the number of timing cycles needed to energize the SN7490 LED.

_____ timing cycles.

Count the number of timing cycles needed to turn off the LED. What does this represent as an output of the IC? _____

_____

_____

Analysis

1. How could the digital timer of this experiment be extended? Name two ways. _____

_____

2. How can the digital timer of this experiment be made variable? _____

_____

3. Would it be advantageous to have a reset switch for an actual digital timer? Why? _____

_____

Activity 87–555 Astable Multivibrators

Name _____ Date _____ Score _____

Objectives

An astable multivibrator repeats itself at the end of each operational cycle. A circuit of this type responds as a clock. The circuit triggers itself into a stage change automatically. Capacitor (C) charges through R_A and R_B and discharges through R_B. The operational cycle is dependent on these components. The charge time $(T_C) = 0.693 \times (R_A + R_B) \times C$. The discharge time $(T_D) = 0.693 \times R_B) \times C$. The total operational time is equal to $T_C + T_D$ or $0.693 \times (R_A + 2R_B) \times C$. Since frequency is a function of time, or $f = 1/T$, then it can be determined by the formula $f = 1.44 \times (R_A + 2R_B) \times C$. The number 1.44 is the reciprocal of 0.693.

In this activity, you will:

1. Construct a clock with a 555 precision timer.

2. Analyze the operation of a 555 clock.

3. Determine the period of operation and frequency of a 555 clock.

Equipment

Multimeter

Power supply—5 Vdc

SE/NE555 IC

Resistors—100 Ω (2), 1 kΩ (2), 1/4 W

Capacitors—100 µF, 1000 µF, 35 V

Silicon diode—Texas Instruments 1N4001

LEDs—20 mA, 2 V (2)

Circuit construction board

Procedure

1. Construct the astable multivibrator clock circuit of Figure 87-1.

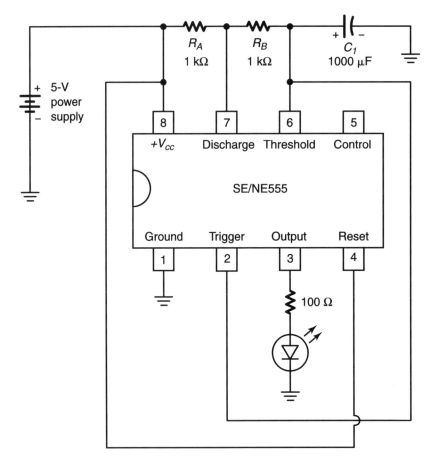

Figure 87-1. Multivibrator clock circuit.

2. Turn on the power supply and test the operation of the circuit. If the circuit is functioning properly, the LED will flash on and off. If this occurs, proceed with Step 3. If it does not, check each circuit connection for possible errors. Then measure the voltage at pins 4 and 8. It should be 5 V. To evaluate the operation of the RC circuit, measure the voltage at pin 6 with respect to ground. This voltage should rise in value to 2/3 of V_{CC} and drop to 1/3 of V_{CC}. This voltage change is necessary to make the circuit change states. If it does not work, turn off the power and measure the resistance and continuity of each component.

3. Determine the charge time (T_C) of the capacitor.

   $T_C =$ _____ s.

4. Determine the discharge time (T_D) of the capacitor.

   $T_D =$ _____ s.

5. Determine the frequency (f) of the circuit.

   $f =$ _____ Hz.

6. The charge time of the capacitor depends on the voltage change that occurs at the threshold terminal (pin 6). This voltage should rise in value to 2/3 of V_{CC} and then fall in value to 1/3 of V_{CC}. With a multimeter, measure the threshold voltage (V_T).

$V_T =$ _____ to _____ V.

7. Calculate the change of the threshold terminal.

Calculated threshold voltage change is _____ to _____ V.

How does this compare with the actual measured voltage change of the circuit? _____

_____

8. Turn off the power supply.

9. Modify the circuit of Figure 87-1 to conform with Figure 87-2. This change is rather minor. The value of C_1 is changed from 1000 µF to 100 µF. In addition to this, a diode is connected across R_B, and a reverse-connected LED is attached to the output. The circuit will show the on and off time of the clock to be approximately the same. C_1 will now charge through only R_A and discharge through R_B. The diode is forward biased during the charging time and reverse biased during the discharge time. As a result, the output will be a symmetrical square wave.

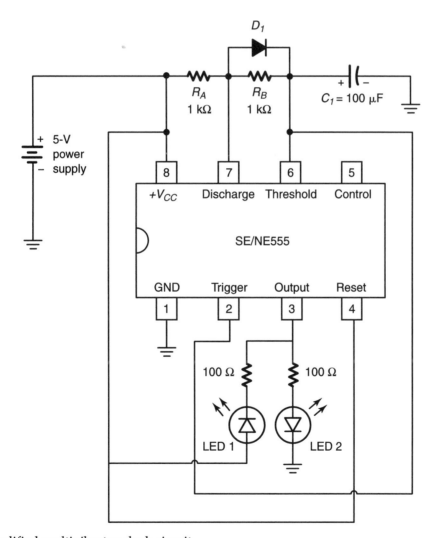

Figure 87-2. Modified multivibrator clock circuit.

10. Turn off the power supply. If the circuit is operating satisfactorily, the LEDs will alternately flash on and off. The on and off time should be equal. How does this circuit compare with Figure 87-1?. _____

11. With a multimeter, measure the voltage across C_1.

Range of voltage change is _____ to _____ V.

12. Calculate the charge time (T_C) and discharge time (T_D) for this circuit. The formula for T_C in this circuit is $0.693 \times R_A \times C$. T_D is the same, $0.693 \times R_B \times C$.

$T_C =$ _____ s; $T_D =$ _____ s.

13. Determine the operational frequency of the circuit in Figure 87-2. In this case, $f = 1.44 \times (R_A + R_B) \times C$.

$f =$ _____ Hz.

14. Turn off the power supply and disconnect the circuit. Return all materials.

Analysis

1. If the clock circuit used in this activity has C changed to 1 μF, how would it alter the frequency of operation? _____

_____

2. If resistors R_A and R_B were changed to 10 kΩ, how would it alter the operation of the circuit?

_____

3. Why does putting a diode across R_A change the output to a symmetrical square wave? _____

_____

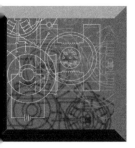

Activity 88–IC Interval Timers

Name _____ Date _____ Score _____

Objectives

The term *interval timer* is used to describe a function where the load is turned on at the start of an operational cycle and kept on for a specific period. At the end of the interval the load is turned off automatically. Mixing operations, photographic exposure timers, and NC machinery operate with this type of timer.

The 555 timer IC will be used in this experiment to demonstrate the interval timing principle. Essentially, this circuit is the same as the delay timer with only the polarity of the load control voltage being changed. You will construct a fixed interval timer and a variable device and calculate the time constant. Through this experiment you will see how the 555 can be used to achieve another unique timing operation.

Equipment

Multimeter

Power supply—5 Vdc

SE/NE555 IC

Resistors—390 Ω, 200 kΩ, 1/8 W

Potentiometer—1 MΩ, 2 W

Capacitor—0.01 µF, 100 Vdc; 50 µF, 25 Vdc

SPST toggle switches (2) (or push buttons)

Light-emitting diode

Circuit mounting board

Procedure

1. Construct the interval timer of Figure 88-1.

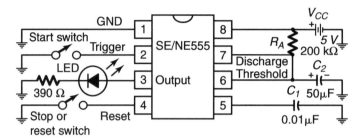

Figure 88-1. IC interval timer.

2. Turn on the 5-V power supply and momentarily turn on the start switch to start the *on* interval. Run several trial tests to verify the on interval.

   Average time is _____ seconds.

3. Calculate the interval with the $T = 1.1RC$ formula.

4. The timing sequence can be reset after it has started by momentarily turning on the reset switch. The cycle can be started again without going through the complete interval.

5. Start the interval, then reset the operation to verify this function.

6. Using the nomograph in Figure 88-2, select a resistor–capacitor combination from any of the spare parts. Estimate the interval, insert R_A and C_2 into the circuit, and test the circuit operation.

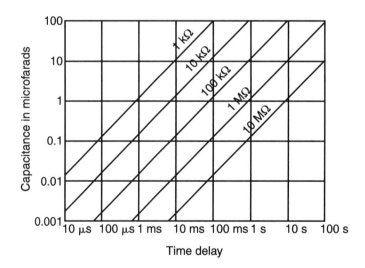

Figure 88-2. RC nomograph.

7. Prepare a multimeter to measure the dc voltage from pin 3 to ground. Start the interval and observe the time. How does it respond? _____

_____

_____

8. Place a 1-MΩ potentiometer in place of R_A. This of course makes the interval variable. Test three positions of the potentiometer at the center, CCW, and CW for different time interval settings. If time permits, you may want to build a variable interval timer control that shows time on a calibrated dial.

9. Turn off the power supply and disconnect the circuit. Return all materials.

Analysis

1. What is the difference in an interval timer compared with a delay timer? _____

_____

2. What are some of the practical limitations of an interval timer of this type? _____

_____

3. What would be an application of this timer? _____

_____

Activity 89–Delay Timer Loads

Name _____ Date _____ Score _____

Objectives

The amount of load current controlled by an IC timer is a very important circuit considera-tion. The SE/NE555 IC timer can be used to control an output circuit of up to 200 mA of current. When larger current values are required, a relay or a driver transistor must be employed. The IC then serves as the driving source that actuates the load control device.

In this experiment, the 555 IC delay timer will first serve as a driver for a relay and then for a transistor load control device. The load can be energized by ac or dc in the relay circuit and only dc in the transistor circuit. The transistor permits load control without moving parts or noise, which is a rather unique advantage over the relay control circuit. Control of ac and large currents are unique advantages of the relay type of system. Through this activity you will be able to see how the 555 timer IC can be used as a signal source for a load control system.

Equipment and Materials

Multimeter

DC power supply—10 V

AC power supply—6.3 V

SE/NE555 IC

Resistors—220 Ω (2), 390 Ω, 200 kΩ, 1/8 W

Potentiometer—1 MΩ, 2 W

Capacitors—0.01 μF, 100 Vdc; 50 μF, 25 Vdc

Transistor—2N3397

Diode—1N4004

Light-emitting diode

Relay—12 V, 125 Ω (Guardian 1335-2C-120D)

SPST toggle switches (2) (or push buttons)

Lamp—No. 47

Circuit mounting board

Procedure

1. Construct the delay timer of Figure 89-1 with a relay.

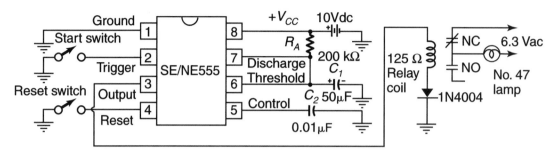

Figure 89-1. Delay timer controlling a relay.

2. The load to be controlled in this circuit is limited to approximately 2 A. This is based upon the contact point rating of the relay. The load lamp of this circuit is controlled by 6 Vac.

3. Initially the load lamp should be on when the ac power is applied.

4. Turn on the dc power source. Test the circuit by momentarily closing the start switch. Then momentarily close the reset switch. Describe the switching action. _____

_____

5. Calculate the delay time with the $T = 1.1R_AC_1$ formula and verify its accuracy.

6. Turn off the ac power source and dc power supply. Prepare the lamp for reverse control by reversing the NO and NC leads to the relay.

7. Apply ac power to the load and dc power to the circuit. Test the operation of the circuit. Describe the operating action of the circuit.

8. Turn off the power to the circuit and to the load lamp. Disconnect the relay from the circuit and replace it with a transistor lamp driver as indicated in Figure 89-2.

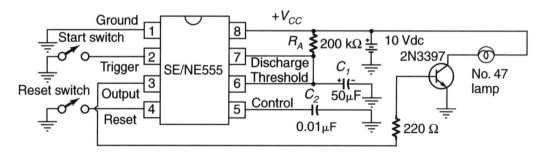

Figure 89-2. Delay timer controlling a lamp.

9. The amount of load current controlled by this circuit is primarily based upon the power rating of the driver transistor. In this circuit, the transistor will control only 1 A of current. The current handling capacity of the circuit could be increased with a higher current rating capacitor. Timers of this type are classified as solid-state electronic devices.

10. Turn off the power and disconnect the circuit. Return all materials.

Analysis

1. What are the advantages of an all solid-state timer over a relay type of delay timer? _____

_____

_____

2. Why is a diode placed in series with the relay coil? _____

_____

_____

3. Are there advantages in using solid-state control to drive a relay? _____

_____

_____

4. Where could a delay off timer be used in an industrial application? _____

_____

_____

5. What would be the problem in controlling the load of the transistor circuit with ac? _____

_____

_____

Activity 90–Mechanical Action Timers

Name _____ Date _____ Score _____

Objectives

Mechanical action timers represent one classification of timing devices that achieve control by entering a mechanical change of some type. Fluid and air flows are typically forced to pass through small openings, called orifices, which restrict their flow. Hydraulic units, air or pneumatic devices, and mercury displacement commonly are used in this type of timer.

In this activity, you will construct a very simple pneumatic mechanical action timer that controls an incandescent lamp. The electrical circuit is switched on and off according to the delay action established by the timer. The device used in this activity is adjustable through a rather narrow range of time. You will test the switching action of the timer and adjust its delay action from one extreme to the other.

As a general rule, the pneumatic action device used in this activity is somewhat less expensive than comparable hydraulic and mercury displacement units. The resulting timing of the pneumatic unit is very similar in nearly all respects to the other units. After working with the pneumatic device one should have a general idea of the capabilities of the other units without actually using the specific devices. All three mechanical action devices are very reliable and have a very long operational life expectancy. The accuracy is generally from 5 to 10% of the set time position. Delay action usually does not exceed 60 minutes.

Equipment and Materials

Multimeter

AC power supply—120 V

Pneumatic timer assembly (Eagle Signal 86-0-A0021)

Incandescent lamp and socket—7.5 W

Isolation transformer—120 Vac

Safety

Be very careful when working with high voltages.

Procedure

1. Prepare the multimeter to measure resistance in the R × 1 range. Measure the continuity between any two of the timer switch contacts. The recommended pneumatic timer for this activity should contain a SPDT switch. Determine the pivot terminal or hinge point of the electrical terminal.

2. Press the push button. If you have selected the correct hinge point of the switch, there should be a break in the continuity. The alternate switch terminal and hinge point should show continuity when the button is released.

3. After a short delay time the switch will change back to its original condition. Make a sketch of the switch terminals in the space below. Point out the hinge or pivot terminal.

4. Connect the electrical part of the circuit shown in Fig. 90-1.

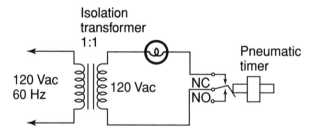

Figure 90-1. Pneumatic mechanical action timer.

5. Depress the push button. Explain the timing action immediately upon release of the push button. Test the circuit and record its initial delay time.

   Initial delay time = _____ seconds.

6. Turn the adjustment needle into the timer until it is firmly seated. Do not force this adjustment or use tools to turn it.

7. Depress the push button and determine the delay time.

   Delay time = _____ seconds.

8. Twist the adjustment needle out of the timer to its full limit of travel. Do not force this adjustment or use tools for this adjustment.

9. Depress the push button and determine the time delay.

   Delay time = _____ seconds.
10. The total range of adjustment is from _____ to _____ seconds.
11. Turn off the power supply and disconnect the circuit. Return all materials.

Analysis

1. Discuss the operation of a mechanical action timer. _____

_____

_____

_____

_____

_____

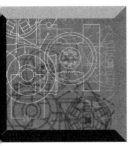

Activity 91–Light-Emitting Diodes

Name _____ Date _____ Score _____

Objectives

In this activity, you will observe the characteristics of a light-emitting diode (LED). These devices are used for numerous applications in industry and elsewhere.

Equipment and Materials

Multimeter

Variable dc power source—0–10 V

Light-emitting diode

Resistor—1 kΩ

Procedure

1. Construct the circuit shown in Figure 91-1.

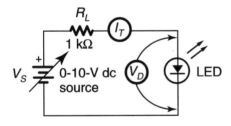

Figure 91-1.   LED test circuit.

2. Before applying power, measure and record the forward and reverse resistance of the LED.

Forward resistance = _____ Ω.

Reverse resistance = _____ Ω.

3. Apply power and complete Figure 91-2.

4. Turn off the power supply and disconnect the circuit. Return all materials.

V_S	V_{R_L}	V_D	I_T	$R_D = V_D / I_T$	Light emitted*
1 V					
2 V					
3 V					
4 V					
5 V					
6 V					
7 V					
8 V					
9 V					
10 V					

*Light as detected by the human eye, i.e., dim, moderate, bright.

Figure 91-2.   Table of LED characteristics. Fill in all values.

Analysis

1. Discuss the operation of an LED. _____

_____

_____

_____

Activity 91—Light-Emitting Diodes

Activity 92–Photovoltaic Cells

Name _____ Date _____ Score _____

Objectives

Many sources of electrical energy exist. These sources include chemical action (batteries, fuel cells), radioactive materials (atomic reactors), heat (thermocouples), pressure (piezoelectric crystals), conductive gases (magnetohydrodynamic generators), electrical generators (moving coils and magnetic fields), and light.

Light is a form of energy that can be converted to electrical energy with very little effort. The most used device to convert light energy into electrical energy is the photovoltaic cell, sometimes called a sun or solar cell. Photovoltaic cells are used extensively in our space program to collect the rays of the sun and convert them into electrical energy. This converted energy is used to power the various circuits that control space satellites, lunar modules, and other space craft.

Photovoltaic cells are usually constructed of two layers of semiconductor material whose electrical characteristics have been altered by the addition of other elements or impurities. When the photovoltaic cell is exposed to light, the two layers interact to cause an excessive number of electrons to exist in one layer. The remaining layer, then, is deficient in electrons. This imbalance in the electrons causes a difference of potential (emf or voltage) to exist between the semiconductive layers. The difference in potential is directly proportional to the amount of light falling upon the cell. This voltage is used to cause current through a load. Thus, light can create electrical energy or power.

In this activity, you will examine the characteristics of the photovoltaic cell as it is used to convert light energy into electrical energy.

Equipment and Materials

- Multimeter
  Variable ac power supply
- Lamp with socket—60 W
- Connecting wires
  Photovoltaic cell

Procedure

1. Prepare your multimeter to measure dc volts. Complete the following connections as illustrated in Figure 92-1.

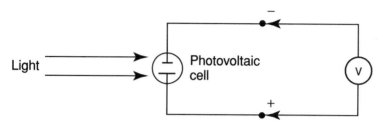

Figure 92-1. Setup for measuring the voltage produced by a photovoltaic cell.

2. Record the voltage output generated by the photovoltaic cell in total darkness and at room light.

$V =$ _____ $V =$ _____
    darkness   room light

3. Connect the 60-watt lamp to the variable ac power supply as shown in Figure 92-2. The power supply must be adjusted to zero and must be turned off.

Figure 92-2. Light source for the photovoltaic cell.

4. Place the 60-watt lamp within 1/2 inch of the surface of the photovoltaic cell. Turn on the variable ac power supply and complete Figure 92-3 by adjusting the ac voltage as indicated. Chart your values in the graph of Figure 92-4.

Power supply voltage (Vac)	Generated cell voltage (mV)
20	
40	
60	
80	
85	
90	
95	
100	
105	
110	
115	
120	

Figure 93-3. Photovoltaic cell outputs.

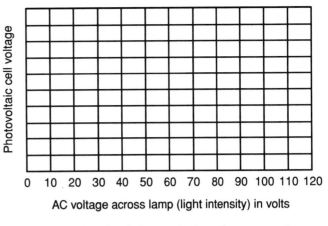

Photovoltaic cell voltage

0 10 20 30 40 50 60 70 80 90 100 110 120

AC voltage across lamp (light intensity) in volts

Figure 93-4. Graph of photovoltaic cell output voltage.

5. Using the graph, what is the relation between light intensity and generated photovoltaic cell voltage? _____

_____

6. Turn off the power supply. Return all materials.

Analysis

1. List five sources of electrical energy. _____

_____

2. Where might photovoltaic cells be used? _____

_____

3. What are some advantages of converting light to electrical energy? _____

_____

_____

Activity 93–Light-Activated Silicon Controlled Rectifier

Name _____ Date _____ Score _____

Objectives

In this activity, you will observe the operational characteristics of a light-activated silicon controlled rectifier (LASCR). The LASCR is a four-layered solid-state device that can be triggered into conduction by focusing light onto its exposed surface. It operates as a conventional SCR since it requires a positive voltage on the gate to cause it to conduct. The major difference between the SCR and LASCR is that the LASCR will conduct with no positive gate voltage as long as a light source is focused onto its surface. The LASCR can be used to activate a larger SCR for control of larger loads.

Equipment and Materials

- Multimeter
- Oscilloscope
- DC power source—6 V
- AC power source—3V
- LASCR
- Resistor—22 Ω
- Potentiometer—1 MΩ
- Lamp—6 V

Procedure

1. Adjust the multimeter to the R × 1000 range. Connect the negative lead to the cathode and the positive lead to the anode of the LASCR. In Figure 93-1, record the anode–cathode resistance with and without light.

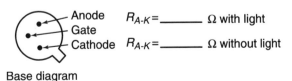

Base diagram

Figure 93-1. LASCR diagram and resistances.

2. Using the meter as the voltage source (connected as described in Step 1) trigger the LASCR with light, without using the gate. Now trigger the LASCR, without light, using the gate and a jumper wire to the positive lead. It should be noted that the LASCR can be triggered with light and no gate connection, with gate potential and no light, or by using a combination of light and gate potential.

3. Construct the circuit shown in Figure 93-2.

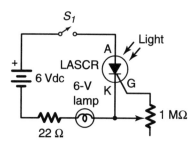

Figure 93-2. A dc LASCR circuit.

4. Place the LASCR in darkness and adjust the 1-MΩ potentiometer to maximum resistance.

5. Close S_1 and adjust the potentiometer until the LASCR activates. Disconnect the potentiometer and measure and record its resistance.

$R_{POT} = $ _____ Ω.

6. Open S_1 and adjust the potentiometer to maximum resistance.

7. Open S_1 and expose the LASCR to normal room light. Adjust the potentiometer until the LASCR activates. Disconnect the potentiometer and measure and record its resistance.

$R_{POT} = $ _____ Ω.

8. Construct the circuit shown in Figure 93-3.

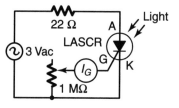

Figure 93-3. An ac LASCR circuit.

9. Adjust the potentiometer until the LASCR activates with normal room light.

10. Connect an oscilloscope across the LASCR (anode to cathode) and record in Figure 93-4 the waveform that appears.

Activity 93—Light-Activated Silicon Controlled Rectifier

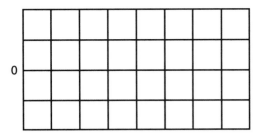

Figure 94-4. Observed LASCR waveform.

11. Using an opaque object, partially shield the LASCR from light until the LASCR is delayed in firing.

12. By altering the position of the opaque object, cause the waveform across the LASCR to appear as illustrated in Figure 93-5. Record the gate current which causes each waveform to be produced.

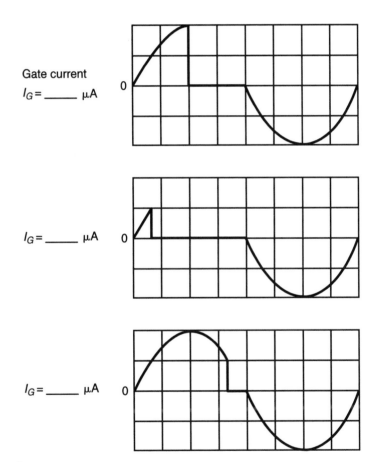

Gate current

$I_G =$ _____ μA

$I_G =$ _____ μA

$I_G =$ _____ μA

Figure 94-5. Observed gate current.

13. Turn off the power supply. Return all materials.

Analysis

1. Discuss the operation of the LASCR. _____

_____

_____

2. From the data of Steps 5 and 7, how do the resistances of the potentiometer compare? Why?

_____

_____

3. From the data obtained in the first circuit constructed, discuss the characteristics of the LASCR with and without light. _____

_____

_____

4. From the data of Step 12, as the conduction angle of the LASCR increases, what happens to gate current? _____

_____

_____

5. Draw a circuit using a LASCR and an SCR that would control a high-current load.

Activity 94–Phototransistors

Name _____ Date _____ Score _____

Objectives

In this activity, you will observe the characteristics of a phototransistor. These devices are ordinarily used to activate other semiconductor devices that are capable of handling greater amounts of current.

Equipment and Materials

- DC power source—5 V
- Phototransistor—GE-X19
- Resistors—1 kΩ, 100 kΩ
- Potentiometer—50 kΩ
- Lamp—60 W

Procedure

1. Using the lead diagram shown in Figure 94-1, measure and record the phototransistor emitter–base (E–B) resistance, collector–base (C–B) resistance, and the emitter–collector (E–C) resistance when forward and reverse biased.

Forward biased   Reverse biased

B C E

Top view

$R_{E-B} =$ _____ $R_{E-B} =$ _____

$R_{E-C} =$ _____ $R_{E-C} =$ _____

$R_{C-B} =$ _____ $R_{C-B} =$ _____

Figure 94-1. Phototransistor lead diagram and resistances.

2. Connect the positive lead of the meter to the collector and the negative lead to the emitter and record this resistance with the transistor in darkness and in normal room light.

Dark resistance = _____ Ω.

Light resistance = _____ Ω.

3. Construct the circuit shown in Figure 94-2.

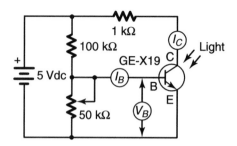

Figure 94-2. Phototransistor test circuit.

4. Adjust the 50-kΩ potentiometer until the collector current is about 3 mA.

5. Place a 60-W lamp about 6 inches from the phototransistor. (Note: Bulb should be adjusted to its maximum brightness.)

6. Complete Figure 94-3 with and without the phototransistor exposed to light.

Light condition	V_B	V_{E-C}	I_B	I_C	$R_{E-C} = V_{E-C}/I_{E-C}$
With light					
Without light					

Figure 94-3. Phototransistor characteristics.

7. Turn off the power supply. Return all materials.

Analysis

1. As light intensity increases, according to the data of Figure 94-1, what happens to:
   a. Base voltage (V_B). _____.
   b. Emitter-collector voltage (V_{E-C}). _____.
   c. Base current (I_B). _____.
   d. Collector current (I_C). _____.
   e. Emitter-collector resistance (R_{E-C}). _____.

2. Discuss the operation of a phototransistor. _____
   _____
   _____

Activity 95–Liquid Level Control

Name _____ Date _____ Score _____

Objectives

Many industrial processes rely upon liquid level measurement. Sometimes variables such as fuel supply are monitored continuously. There are several techniques that can be used to measure liquid level.

The measurement of liquid level is easy to accomplish by using transducers. Level changes result in the displacement of the top surface of the liquid. Many types of transducers are used to measure liquid level. Resistive transducers can be used to measure the level of a conductive solution. A capacitive transducer can be used along with a movable plate whose position is determined by the level of the liquid. Photoelectric methods, radioactive methods, and ultrasonic methods can also be employed.

A simple type of level controller is the ball float system. This system uses a ball float to operate a lever. The lever is connected to a valve that regulates liquid flow rate. Chemical industries commonly use differential pressure controllers to control the level of volatile liquids. The liquid pressure is proportional to its level in the enclosed container.

Level control can be accomplished by placing a light source at the same height above a conveyor line as the desired fill level of a container. In the illustration of Figure 95-1, containers on a conveyor line are positioned under a liquid dispenser. When the container is in position, the actuator causes the dispenser to allow liquid to pass into the container. When the liquid reaches the level of the light source, the light beam is interrupted. With no light striking its surface, the detector will cause a relay to activate. The activated relay will, in turn, cause the actuator on the dispenser to close. When another container is in position, the actuator will open once more. This liquid level control system ensures a uniform level of liquid in each container.

In this activity, you will construct and test a photoelectric circuit that can be used to measure liquid level.

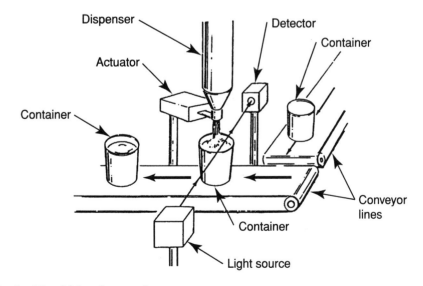

Figure 95-1. Method of liquid level control.

Equipment and Materials

- Multimeter
- Variable dc power supply
- AC power supply—120 V
- Light-dependent resistor (GE-X6 or equivalent)
- Container—glass or plastic
- Relay—12 V, 1250 Ω (Guardian 1335-2C-120D or equivalent)
- Lamp with holder—7 W
- Lamp with holder—60 W
- SPST switch

Safety

Be very careful when working with high voltages.

Procedure

1. Construct the photoelectric liquid level control of Figure 95-2.

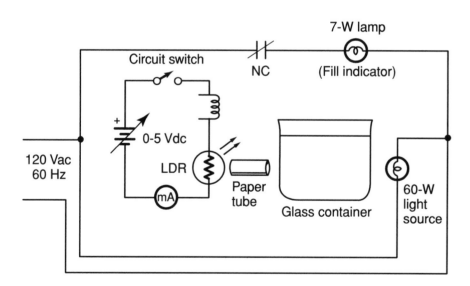

Figure 95-2. Photoelectric liquid level control.

2. Plug the light source into a 120 Vac power outlet and close the circuit switch. Position the light source near the bottom of the glass container before it is filled with water. Slide the paper tube around the light-dependent resistor (LDR) and place the open end against the container.

3. If the circuit is operating properly, the fill indicator lamp will light when the relay is actuated.

4. Adjust the dc source to alter the sensitivity of the circuit. Do not exceed 15 Vdc. The circuit should be able to detect a pencil passing in front of the paper tube.

5. Carefully fill the container with water until the indicator turns off. Avoid pouring water directly in front of the tube window area. You may need to try several trial runs to get the sensitivity to a level where the circuit will respond properly.

6. After the sensitivity has been adjusted, drain or siphon water from the container until the fill indicator is actuated again.

7. Test the liquid level control circuit two or three times to ensure that it operates properly.

8. Turn off the power. Return all materials.

Analysis

1. Discuss the operation of the circuit used in this activity. _____

_____

_____

_____

_____

2. What are some other types of circuits that could be used for liquid level control? _____

_____

_____

_____

_____

Activity 96–Binary Numbers

Name _____ Date _____ Score _____

Objectives

Practically all of the electronic digital systems in operation employ binary numbers. This type of numbering system uses the number 2 as its base or radix. The two conditions or states of this system can be on or off, voltage or no voltage, or the number designations of 1 and 0. The process of changing from one state to the other can be easily achieved with switches, solid-state devices, or integrated circuits.

In this activity, a number of practice problems will be provided so that you can become more proficient in the use of the binary numbering system. The first problems deal with conversion of base 10 numbers to binary equivalents. Then binary numbers are converted to base 10 equivalents. The latter part of the activity is then devoted to binary coded decimal (BCD) conversions.

The practice problems provided in this activity are primarily designed to improve your ability to manipulate binary numbers. These numbers play an important role in the analysis of digital electronic system circuitry.

Procedure

1. Change the following decimal numbers (base 10) to equivalent binary numbers (base 2). Show all of the necessary steps needed to arrive at the final answer.

   $12_{10} =$ _____$_2$

   $37_{10} =$ _____$_2$

   $62_{10} =$ _____$_2$

   $59_{10} =$ _____$_2$

   $60_{10} =$ _____$_2$

   $15.5_{10} =$ _____$_2$

   $22_{10} =$ _____$_2$

   $41_{10} =$ _____$_2$

   $47_{10} =$ _____$_2$

   $83_{10} =$ _____$_2$

   $102_{10} =$ _____$_2$

   $9.75_{10} =$ _____$_2$

2. Change the following binary numbers to their decimal equivalents.

$1101._2 = \underline{\hspace{2cm}}_{10}$

$101111._2 = \underline{\hspace{2cm}}_{10}$

$100010._2 = \underline{\hspace{2cm}}_{10}$

$10.1_2 = \underline{\hspace{2cm}}_{10}$

$11.11_2 = \underline{\hspace{2cm}}_{10}$

$11100._2 = \underline{\hspace{2cm}}_{10}$

$11010._2 = \underline{\hspace{2cm}}_{10}$

$11000._2 = \underline{\hspace{2cm}}_{10}$

$101.01_2 = \underline{\hspace{2cm}}_{10}$

$110.111_2 = \underline{\hspace{2cm}}_{10}$

3. Express the following decimal numbers in binary coded decimal form.

$362._{10} = \underline{\hspace{2cm}}_{BCD}$

$683._{10} = \underline{\hspace{2cm}}_{BCD}$

$741._{10} = \underline{\hspace{2cm}}_{BCD}$

$405._{10} = \underline{\hspace{2cm}}_{BCD}$

$1248._{10} = \underline{\hspace{2cm}}_{BCD}$

$329._{10} = \underline{\hspace{2cm}}_{BCD}$

$197._{10} = \underline{\hspace{2cm}}_{BCD}$

$29._{10} = \underline{\hspace{2cm}}_{BCD}$

$758._{10} = \underline{\hspace{2cm}}_{BCD}$

4. Change the following BCD expressions into an equivalent decimal, or base 10, value.

$0101\text{-}0111\text{-}1001_{BCD} = \underline{\hspace{2cm}}_{10}$

$0011\text{-}0010\text{-}0100_{BCD} = \underline{\hspace{2cm}}_{10}$

$1001\text{-}0111\text{-}0110_{BCD} = \underline{\hspace{2cm}}_{10}$

$1001\text{-}0010\text{-}0001_{BCD} = \underline{\hspace{2cm}}_{10}$

$0101\text{-}0110\text{-}0011_{BCD} = \underline{\hspace{2cm}}_{10}$

$1000\text{-}0111\text{-}0000_{BCD} = \underline{\hspace{2cm}}_{10}$

5. Perform the operations indicated.

$101._{10} = \underline{\hspace{2cm}}_{2}$

$1101101.1_2 = \underline{\hspace{2cm}}_{10}$

$67._{10} = \underline{\hspace{2cm}}_{2}$

$110.11_2 = \underline{\hspace{2cm}}_{10}$

Analysis

1. In a positive logic system, +5 V would be assigned _____ value and 0 V would be assigned _____ for a binary system.

2. What would be an advantage of expressing a decimal number in BCD form? _____

   _____

   _____

   _____

3. Why is the binary system particularly well suited for electronic switching applications? _____

   _____

   _____

   _____

4. In a BCD expression, why is the largest number in any one position 9 when a number as large as 1111 or $15._{10}$ can be displayed by four binary numbers? _____

   _____

   _____

   _____

5. What is the radix of the decimal system? What is the radix of the binary system? _____

   _____

   _____

   _____

6. What is meant by the term *place value* in the description of a numbering system? _____

   _____

   _____

   _____

Activity 97—Octal and Hexadecimal Numbers

Name _____ Date _____ Score _____

Objectives

Octal and hexadecimal numbers are commonly used in digital systems to process large numbers. These two systems are also very compatible with the binary system. Conversions from binary to octal or hexadecimal are accomplished by simply grouping values together in threes or fours. The octal system has a radix of with place value assignments of 1s, 8s, 64s, 512s, 4096s, 32,768s, etc. The hexadecimal system is similar, but has a radix of sixteen. Place value assignments are 1s, 16s, 256s, 4096s, 65,536s, etc. Larger number values can obviously be represented by the hexadecimal system in fewer places.

In this activity, you will practice number system conversions. Binary, octal, hexadecimal, and decimal numbers all play an important role in digital systems. You will become more familiar with these conversion processes after working the problems in this activity.

Procedure

1. Change the given octal numbers to decimal equivalents.

$123._8$ = _____$_{10}$

$4062._8$ = _____$_{10}$

$712._8$ = _____$_{10}$

$176._8$ = _____$_{10}$

$3120._8$ = _____$_{10}$

$643._8$ = _____$_{10}$

$541._8$ = _____$_{10}$

$674._8$ = _____$_{10}$

$1026._8$ = _____$_{10}$

2. Change the given octal numbers to binary equivalents.

$125._8$ = _____$_2$

$73._8$ = _____$_2$

$36._8$ = _____$_2$

$136._8$ = _____$_2$

$46._8$ = _____$_2$

$54._8 = $ _____ $_2$

$412._8 = $ _____ $_2$

$67._8 = $ _____ $_2$

$107._8 = $ _____ $_2$

3. Change the given decimal numbers to octal (base 8) equivalents.

$4093._{10} = $ _____ $_8$

$1017._{10} = $ _____ $_8$

$3699._{10} = $ _____ $_8$

$5012._{10} = $ _____ $_8$

$4876._{10} = $ _____ $_8$

$4162._{10} = $ _____ $_8$

$3316._{10} = $ _____ $_8$

$1491._{10} = $ _____ $_8$

$1016._{10} = $ _____ $_8$

4. Change the given binary numbers to octal equivalents.

$100,110,100._2 = $ _____ $_8$

$101,111,111._2 = $ _____ $_8$

$110,110._2 = $ _____ $_8$

$1,001,100._2 = $ _____ $_8$

$1,001,000._2 = $ _____ $_8$

$11,111,010._2 = $ _____ $_8$

5. Convert the given hexadecimal (base 16) numbers to decimal equivalents.

$12A3._{16} = $ _____ $_{10}$

$1BF._{16} = $ _____ $_{10}$

$217C._{16} = $ _____ $_{10}$

$1011B._{16} = $ _____ $_{10}$

$2BCD._{16} = $ _____ $_{10}$

$14BD._{16} = $ _____ $_{10}$

$1A._{16} = $ _____ $_{10}$

$15E._{16} = $ _____ $_{10}$

$BCD._{16} = $ _____ $_{10}$

6. Convert the given hexadecimal numbers to equivalent binary numbers.

$1479._{16} = $ _____ $_2$

$10BA._{16} = $ _____ $_2$

$144E._{16} = $ _____ $_2$

$13AB._{16} = $ _____ $_2$

$123A._{16} = $ _____$_2$

$129A._{16} = $ _____$_2$

$10AD._{16} = $ _____$_2$

$BDE._{16} = $ _____$_2$

$157B._{16} = $ _____$_2$

7. Convert the given decimal numbers to hexadecimal equivalents.

$4170._{10} = $ _____$_{16}$

$3159._{10} = $ _____$_{16}$

$4153._{10} = $ _____$_{16}$

$5690._{10} = $ _____$_{16}$

$2137._{10} = $ _____$_{16}$

$1176._{10} = $ _____$_{16}$

$4910._{10} = $ _____$_{16}$

$1210._{10} = $ _____$_{16}$

8. Convert the given binary numbers to hexadecimal equivalents.

$1\ 0110\ 1111\ 0100._2 = $ _____$_{16}$

$1\ 1011\ 1000\ 1110._2 = $ _____$_{16}$

$101\ 1011\ 0111._2 = $ _____$_{16}$

$1001\ 1111\ 1101._2 = $ _____$_{16}$

$1\ 0011\ 0110._2 = $ _____$_{16}$

$101\ 1111\ 0110._2 = $ _____$_{16}$

Analysis

1. Why are octal and hexadecimal number systems commonly used in digital systems? _____

   _____

   _____

2. What place values does an octal number with five digits have? _____

   _____

3. What place values does a hexadecimal number have with five digits? _____

   _____

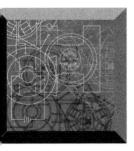

Activity 98–Digital Logic Functions

Name _____ Date _____ Score _____

Objectives

Control of a digital system is achieved by a variety of logic gates. Each gate has a particular output response to a combination of input signals. Binary data is represented by 1s and 0s. Multiple-input gates of the AND and OR variety and a single input/output NOT gate represent the three primary logic functions of a digital system. An understanding of these logic functions is essential when analyzing the operation of a digital system.

In this activity, each basic logic gate will be constructed with a single integrated circuit. Truth tables are then developed to show the relationship between the input alternatives and the corresponding output. Through this activity you will gain experience in using the basic logic gates and see how the logic functions are achieved electronically.

Equipment and Materials

Multimeter

DC power supply—5 V

SN7408 IC

SN7411 IC

SN7432 IC

SN7404 IC

Resistors—470 Ω, 1/8 W (4)

Light-emitting diodes (4)

SPST toggle switch

SPDT toggle switches (3)

IC breadboard construction unit

Procedure

Part A: Gate Testing with the SN7408 and SN7411

1. Connect the logic gate test circuit of Figure 98-1.

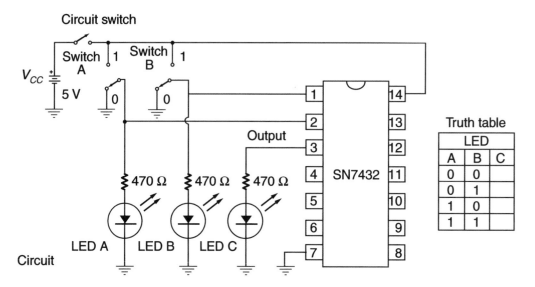

Figure 98-1. SN7408 IC test circuit and truth table.

2. Before closing the circuit switch, turn on the power supply and adjust it to 5 V.

3. Turn on the circuit switch and test the IC gate according to the alternatives listed in the truth table of Figure 98-1.

4. Prepare a multimeter to measure voltage and determine the input and output voltages corresponding to 1 and 0.

   A 1 = _____ V, while a 0 = _____ V.

5. There are three other gates included in this IC chip. Pins 4 and 5 are the input and pin 6 is the output of gate 2. Pins 9 and 10 are the input and pin 8 is the output of gate 3. Pins 13 and 12 are the input and pin 11 is the output of gate 4. Test these gates to see if they are working properly.

6. Make a gate symbol drawing on the blank IC dual in-line package layout of Figure 98-1 for the SN7408.

7. Open the circuit switch and remove the SN7408. In its place connect an SN7411. Be certain that pin 14 is connected to the +5V V_{CC} source and pin 7 is connected to ground. Connect the remainder of the circuit indicated in Figure 98-2.

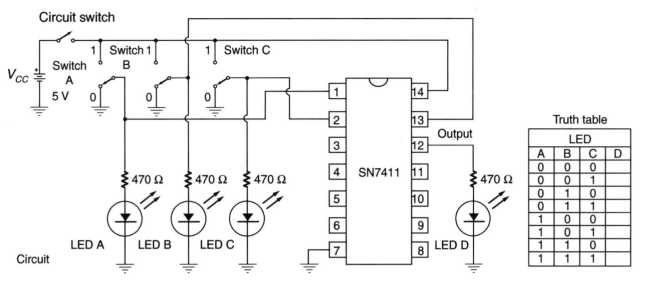

Figure 98-2. SN7411 IC test circuit and truth table.

Truth table

	LED		
A	B	C	D
0	0	0	
0	0	1	
0	1	0	
0	1	1	
1	0	0	
1	0	1	
1	1	0	
1	1	1	

8. Close the circuit switch and test the SN7411. Complete the truth table for the listed input alternatives.

9. Open the circuit switch and disconnect the IC.

Part B: Gate Testing with the SN7432

1. Construct the logic gate test circuit which is shown in Figure 98-3.

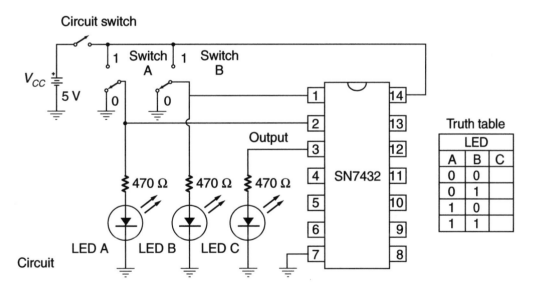

Figure 98-3. SN7432 IC test circuit and truth table.

Truth table

	LED	
A	B	C
0	0	
0	1	
1	0	
1	1	

2. Before closing the circuit switch, turn on the power supply and adjust it to 5 V.

3. Turn on the circuit switch and test the IC gate according to those alternatives that are listed in the truth table of Figure 98-3.

4. What logic gate function is achieved by this gate? _____.

5. There are three other gates included in this IC chip: pins 4 and 5 are inputs and pin 6 is an output; pins 9 and 10 are inputs and pin 8 is an output; and pins 12 and 13 are inputs and pin 11 is an output. Test these gates to see if they are working properly.

6. Make a gate symbol drawing on the blank IC dual in-line package layout of Figure 98-4 showing the input output connections for the SN7432.

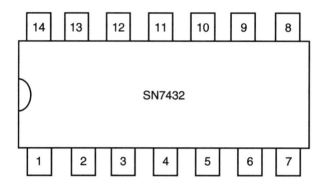

Figure 98-4. SN7432 IC pin connections.

7. Open the circuit switch and disconnect the circuit.

Part C: Gate Testing with the SN7404

1. Connect the logic gate test circuit of Figure 98-5.

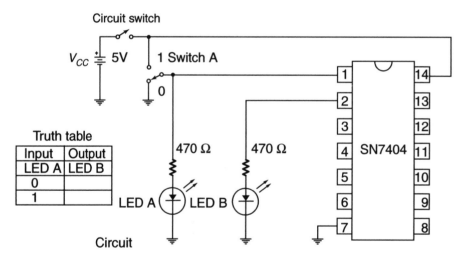

Figure 98-5. SN7404 IC test circuit and truth table.

2. Before closing the circuit switch, turn on the power supply and adjust it to 5 V.

3. Close the circuit switch and test the IC gate according to the alternate input conditions listed in the truth table of Figure 98-5. Record the respective output for each of the inputs.

4. What logic function is achieved by this gate? _____.

5. Test the remaining five gates at pins 3–4, 5–6, 9–8, 11–10, and 13–12. (The pin number designation listed first is the input, with the last number indicating the output.)

6. Make a gate symbol drawing on the blank dual inline package layout of Figure 98-6 showing the input-output connections for each gate of the SN7404.

Activity 98—Digital Logic Functions

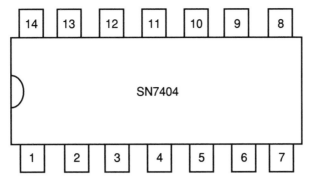

Figure 98-6. SN7404 IC pin connections.

7. Open the circuit switch and connect the output of gate 1 to the input of gate 2. Connect the output indicating LED to pin 4.

8. Close the circuit switch and test the logic circuit. What function does this indicate? _____.

9. Open the circuit switch and disconnect the circuit. Return all parts.

Analysis

1. What logic functions are achieved by the gates studied in Part A? _____

_____

_____

2. What mathematical functions are achieved by the gates studied in Part A? _____

_____

_____

3. What are the symbolic representations of the gates studied in Part A? _____

_____

_____

4. Prepare a statement that describes the operation of the gates studied in Part A. _____

_____

_____

5. What mathematical function is achieved by the gate studied in Part B? _____

_____

_____

6. What are the two logic symbols that are used to represent the circuit constructed in Part B of this activity? _____

_____

_____

7. Prepare a statement that describes the operation of the gate studied in Part B. _____

_____

8. What mathematical function is achieved by the logic gate studied in Part C? _____

_____

_____

9. What is meant by the terms *negation* and *double negation*? _____

_____

_____

10. What would be the resulting output of a 1 applied to three of the gates studied in Part C of this activity connected in series? _____

_____

Activity 99–Combination Logic Rates

Name _____ Date _____ Score _____

Objectives

When either an AND gate or an OR gate is connected to a NOT gate, two additional logic functions are achieved. A NOT–AND, or NAND, is one type of combination logic gate. The NOT–OR, or NOR, gate is representative of the second type of combinational logic gate achieved.

NOR, NAND, and NOT gates are often considered to be universal building blocks in digital systems. With these gates, it is possible to build four logic gates plus the original function. A person working with digital systems should be very familiar with the universal building block principle of combinational logic gates.

In this activity, you will investigate the NAND, NOR, and NOT gates in combinational logic gate construction operations. Only gates that have an inverting capability can be used in this building block technique.

Equipment and Materials

- Multimeter
- DC power supply—0-5 V, 1 A
- SN7400 IC
- SN7402 IC
- SN7404 IC
- SPST toggle switch
- SPDT toggle switches (2)
- Resistors—470 Ω, 1/8 W (3)
- Light-emitting diodes (3)
- IC circuit construction board

Procedure

Part A: NAND Logic

1. Using the SN7400 quad NAND of Figure 99-1, build the IC test circuit and check gate 1.

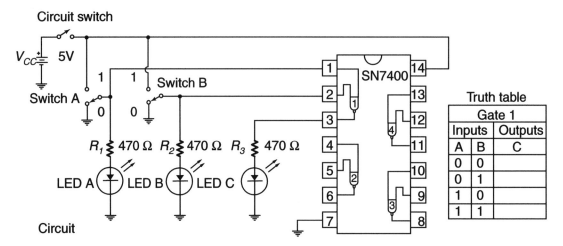

Figure 99-1. SN7400 IC test circuit and truth table.

2. Record the 1s and 0s of the output in the truth table for each of the input alternatives.

3. With a multimeter, measure and record voltage values represented by the 1s and 0s of this gate.

   A 1 = _____ V, while a 0 = _____ V.

4. Test the other three gates to make certain that they are functioning properly.

5. To build a NOT gate, remove the B circuit from pin 2 and connect pins 1 and 2 together. Use switch A as the input and test the output of the gate.

6. Connect NAND gate 1 as instructed in Step 1. Connect the output of gate 1 to pins 4 and 5 of NAND 2. Connect the output LED to pin 6. What function does this combination logic gate achieve? _____

7. Test the circuit to verify your theory. Did it work as you predicted? _____

8. Connect the combinational logic circuit of Figure 99-2. Complete the truth table showing the outputs at the indicated points.

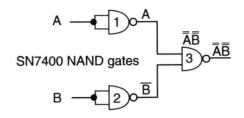

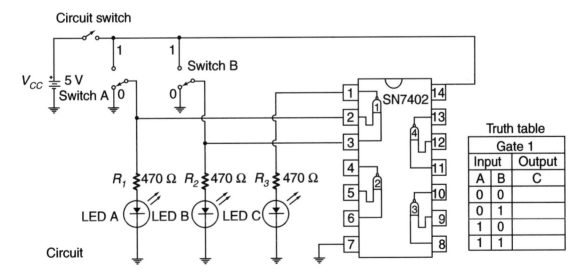

Input		Expression			Output
A	B	$\overline{A}$	$\overline{B}$	$\overline{AB}$	$\overline{\overline{AB}}$
0	0				
0	1				
1	0				
1	1				

Figure 99-2. Combination NAND gate circuit and truth table.

9. According to the truth table, what gate function is achieved by this combination logic gate? Test the gate to verify your prediction. Did it perform the function that you expected? _____

_____

10. Open the circuit switch and disconnect the circuit.

Part B: NOR Logic

1. Using an SN7402 quad NOR gate, Figure 99-3, build the IC test circuit and check gate 1.

Figure 99-3. SN7402 IC test circuit and truth table.

Gate 1		
Input		Output
A	B	C
0	0	
0	1	
1	0	
1	1	

2. Record the 1 and 0 outputs for the input alternatives listed in the truth table in Figure 99-3.

3. With a multimeter, measure the voltage values represented by the 1s and 0s of this gate.

A 1 = _____ V, while a 0 = _____ V.

4. Test the other three gates to make certain that they functioning properly.

5. To build a NOT gate, disconnect the lead to pin 3 and then connect pin 2 to pin 3. Use switch A as the input and test the output of the gate.

6. Connect NOR gate 1 as instructed in Step 1. Connect the output of NOR 1 to the input of NOR 2, with NOR 2 connected as a NOT gate. Connect the output LED to pin 4. What combination logic is achieved by this configuration? _____

7. Test the circuit to verify your theory. Did it work as you predicted? _____

8. Connect the combinational logic circuit of Figure 99-4 using the SN7402.

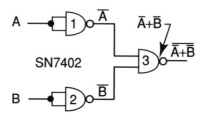

Inputs		Expression			Output
A	B	$\overline{A}$	$\overline{B}$	$\overline{A}+\overline{B}$	$\overline{\overline{A}+\overline{B}}$
0	0				
0	1				
1	0				
1	1				

Figure 99-4. Combination NOR gate circuit and truth table.

9. Complete the truth table in Figure 99-4 for outputs at the designated points. Then test the combinational logic circuit. Does it do what you predicted by truth table? _____

10. Open the circuit power switch and connect NOR gate 4 as a NOT gate. Attach the output in Figure 99-4 to the input of NOR 4. Connect the LED to the output of NOR 4. Invert the last stage of the truth table of Figure 99-4 and predict the type of gate achieved by your modified circuit.

11. Close the circuit switch and test your prediction. Was it correct? _____

12. Open the circuit switch and disconnect the circuit.

Part C: NOT Logic

1. Using the SN7404 hex-inverter IC of Figure 99-5 build the IC test circuit and check gate 1.

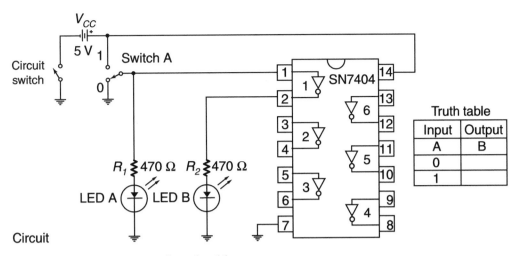

Figure 99-5. SN7404 IC test circuit and truth table.

Truth table	
Input	Output
A	B
0	
1	

2. Turn on the circuit switch and record the 1s and 0s of the output in the truth table of Figure 99-5 for the input alternatives.

3. With a multimeter, measure the voltage values of a representative 1 and 0 of this gate.

   A 1 = _____V, while a 0 = _____ V.

4. Test the other five gates to verify that they are operating properly.

5. Open the circuit switch and combine gates 1 and 2 as indicated in Figure 99-6.

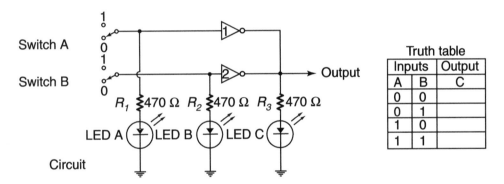

Figure 99-6. Combination NOT gate circuit and truth table.

Truth table		
Inputs		Output
A	B	C
0	0	
0	1	
1	0	
1	1	

6. Close the circuit switch and test the combination gate. Record the 1 and 0 outputs for each of the input alternatives of the truth table in Figure 9-6. What logic function does this achieve?

7. Open the circuit switch and connect the output of the circuit in Figure 99-6 to the input of gate 3. Connect the LED-resistor to pin 6 and test the circuit. What logic function does it achieve?

8. Open the circuit switch and combine gates of the inverter to form the circuit of Figure 99-7.

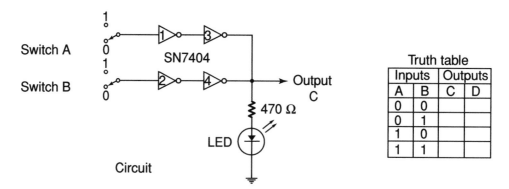

Figure 99-7. Cascaded double-inverter circuit.

9. Close the circuit switch and test the circuit. Record the 1 and 0 outputs of C in the truth table of Figure 99-7 for the given input alternatives. What gate function does this achieve? _____

10. Open the circuit switch and add the output of Figure 99-7 to the input of gate 5. Move the LED to the output of gate 5. What gate function do you predict the output will demonstrate? ____

11. Close the circuit switch and test the circuit. Record your findings at the D output of the truth table in Figure 99-7. Was your prediction correct? _____

12. Open the circuit switch and disconnect the components. Return all parts.

Analysis

1. Complete Figure 99-8 showing the outputs for each logic gate listed.

Inputs		Outputs			
A	B	NAND	AND	OR	NOR
0	0				
0	1				
1	0				
1	1				

Figure 99-8.

2. Make a sketch of the five logic circuit combinations accomplished with the SN7402. Show the actual logic gate combinations used.

Activity 99—Combination Logic Gates

3. In any of the combined inverter gate combinations, why is the 0 considered a predominate factor? _____

_____

_____

4. Make a sketch of the five logic functions achieved by the SN7404. _____

Activity 100–R-S Flip-Flops

Name _____ Date _____ Score _____

Objectives

A bistable latch is often used to explain the operation of a flip-flop. The terms latch and flip-flop can be used interchangeably. The latch used here employs a pair of cross-connected two input logic gates. Either NOR or NAND gates can be used to accomplish this function. The logic gates of this configuration permit the circuit to remain in one state or the other depending on whether it is set or reset. R-S flip-flops are used to debounce switches and to store data temporarily. R-S flip-flops are not available today on IC chips.

In this activity, you will:

1. Construct R-S flip-flops with logic gates.

2. Develop a truth table that shows the operation of an R-S flip-flop.

3. Evaluate the operation of an R-S flip-flop or latch.

Equipment

- Power supply—0-5 Vdc

- SN7400 quad two-input NAND gate

- SN7402 quad two-input NOR gate

- Resistors—100 Ω (2), 1 kΩ (2), 1/4 W

- LEDs—20 mA, 2 V

- SPDT slide switches (2)

- Circuit construction board

Procedure

1. Connect the R-S flip-flop using two cross-connected NOR gates of an SN7402 in Figure 100-1. Any two of the four NOR gates can be used. The pin-out of gates 1 and 2 is shown. Pin 7 is connected to ground and pin 14 to + 5 V to energize the IC.

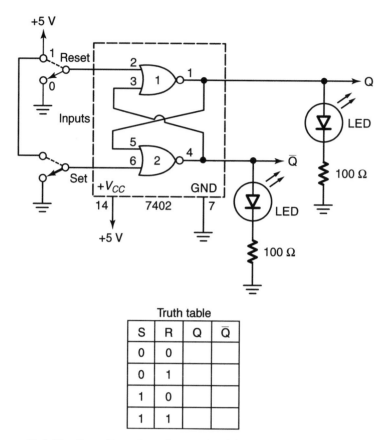

Figure 100-1. NOR gate R-S flip-flop. Complete the truth table.

Truth table

S	R	Q	Q̄
0	0		
0	1		
1	0		
1	1		

2. Turn on the 5-V dc source and place the set and reset switches in the ground position. This represents the 0-V, or low-level, input. As a rule, the gate energized first causes its LED to light. This output is generally considered to be unpredictable. Note the behavior of the LEDs in the truth table for this condition of operation. An on LED is considered to be a 1 and an off LED indicates 0.

3. Change the operational states of the input to conform with the remaining three steps of the truth table. Record the status of the output in the truth table.

4. Alter the circuit of Figure 100-1 to conform with the circuit of Figure 100-2. In this case, only the input needs to be changed. With the switch in position 1, record the status of Q and Q̄ in the truth table. Change the switch to position 2. Record the status of Q and Q̄.

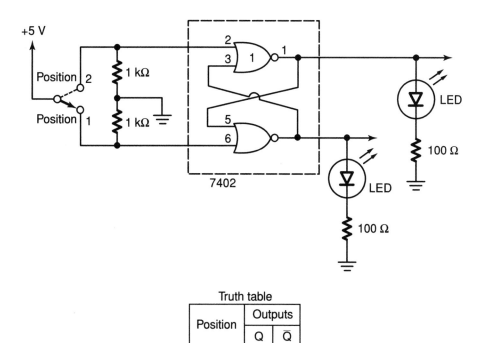

Truth table

Position	Outputs	
	Q	$\overline{Q}$
1		
2		

Figure 100-2. NOR gate circuit. Complete the truth table.

5. Turn off the power supply and remove the SN7402 from the circuit construction board. Replace it with an SN7400. The dc supply voltage is applied to the same pins of this IC. The other pins of the NAND gates are connected in a different manner. Connect the circuit according to Figure 100-3.

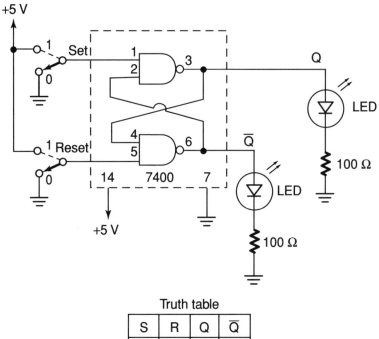

Truth table

S	R	Q	Q̄
0	0		
0	1		
1	0		
1	1		

Figure 100-3. NAND gate R-S flip-flop. Complete the truth table.

6. Turn on the power supply and place the set and reset switches in the ground or 0 position. Note the behavior of Q and Q̄. Record your findings in the truth table for this condition of operation.

7. Change the operational state of the inputs to conform with the other three steps of the truth table. Record the status of Q and Q̄ for each input step.

8. Alter the circuit of Figure 100-3 to conform with the circuit of Figure 100-4. Only the input of the circuit is changed. With the switch in position 1, record the status of Q and Q̄ in the truth table. Change the switch to position 2. Record the status of Q and Q̄ in the truth table.

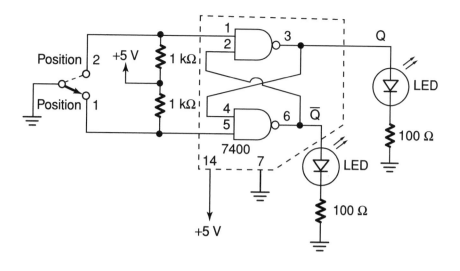

Truth table

Position	Output	
	Q	$\overline{Q}$
1	0	1
2	1	0

Figure 100-4. NAND gate circuit. Complete the truth table.

9. Disconnect the circuit. Return all materials.

Analysis

1. Explain what it takes to set an RS flip-flop constructed with NOR gates. _____

_____

_____

2. Explain what it takes to set an RS flip-flop constructed with NAND gates. _____

_____

_____

Activity 101–Binary Counter ICs

Name _____ Date _____ Score _____

Objectives

The 74193 is a medium-scale-integration (MSI) integrated circuit that is designed to achieve binary counting. This chip has four J-K flip-flops and is considered to be a synchronous up–down 4-bit binary counter. It can be connected as a ripple counter or as a parallel-loaded synchronous counter. MSI chips of this type permit a large number of components to be fabricated into functional devices. Counting operations are generally achieved by chips of this type.

In this activity, you will:

1. Connect and operate a binary counter with a single IC.

2. Connect a binary counter so that it will count up or down.

3. Demonstrate the operation of a binary counter.

Equipment and Materials

- Multimeter

- DC power supply—5 V

  74193 synchronous binary up–down counter

- LEDs—20 mA, 2 V (4)

- Resistors—100 Ω, 1/4 W (4)

  Circuit construction board

- Clock circuit—Any one of the following:

  555 timer (1), 10-kΩ resistors (2), 1N4001 diode (1), 100-mF, 35-V capacitor (1)

  Square wave generator

  Digital trainer with a built-in clock

Procedure

1. Connect the circuit in Figure 101-1. Refer to the pin connections of the 74193. Make connections as follows:

   Pins 15, 1, 10, and 9 (preset inputs)—open

   Pin 12 (carry out)—open

   Pin 13 (borrow out)—open

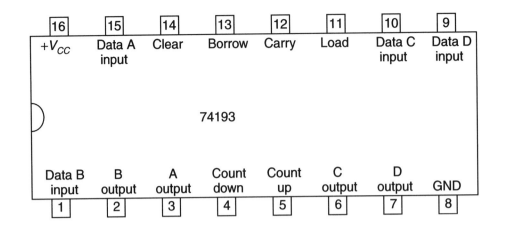

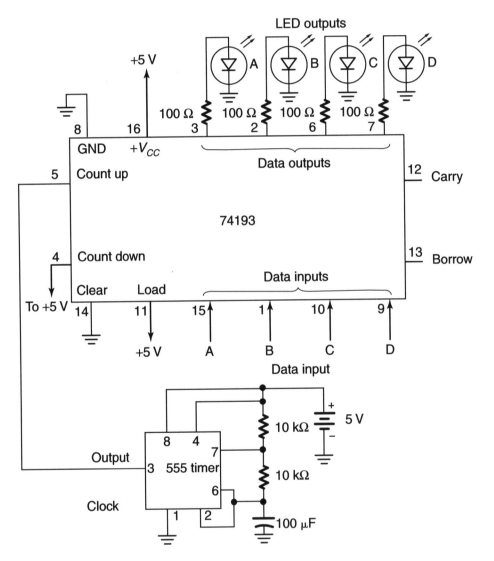

Figure 101-1. Pin layout and circuit setup for binary counter.

Pins 2, 3, 6, and 7 (outputs)—to LEDs

Pin 11 (load)—to +5 V

Pin 16 (+ V_{CC})—to +5 V

Pin 14 (Clear)—to ground

Pin 8 (ground)—to ground

Pin 4 (down)—to +5 V

2. Energize the circuit. If the output is not 0000_2, reset the circuit by momentarily disconnecting pin 14.

3. Connect the clock input to pin 5. This should start the circuit counting. If it does not count, check the output of the clock with a multimeter to see if it is changing states. Recheck the pin connections of the 74193 to see if they conform with the preceding connections.

4. The counting sequence is (up, down).

5. Momentarily disconnect the clock and clear the counter to 0000.

6. Connect the clock to pin 5. Describe the counting sequence in binary numbers. _____

_____

7. Turn off the power supply and disconnect the clock from pin 5 and the +5-V supply from pin 4 of the counter. Connect pin 5 to the +5-V supply and pin 4 to the clock.

8. Energize the circuit and note the counting sequence. This represents an (up, down) counter.

9. Describe the counting sequence in binary numbers. _____

_____

10. Turn off the power supply. The counter circuit and clock should be retained for the next activity.

Analysis

1. What does it take to change the counting direction of a 74193? _____

_____

2. Why is the circuit of this activity considered to be a ripple counter? _____

_____

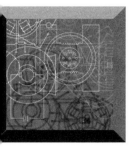

Activity 102–Binary-Coded-Decimal Counters

Name _____ Date _____ Score _____

Objectives

One of the most significant operations of a digital system is counting. As a general rule, counting is achieved by binary numbers because of the ease by which two-state data can be controlled. Since most people are familiar with the decimal, or base 10, counting system, digital counters must make the transition from binary to decimal to be useful. Binary-coded-decimal (BCD) counters are used to display decimal information in a binary code. Only the numbers 0 through 9 are displayed in binary form. Counts beyond 1001_2, or 9_{10}, are invalid and are not used.

In this activity, a BCD counter is constructed with a single IC. The counter is manually triggered through its operation with a debounced switch. This permits circuit changes to be made while observing the response of the output. A clock signal is then applied to the counter to show how it responds automatically.

The 7490 BCD counter is used in this activity. It employs four flip-flops and a gate circuit to alter its count. The counter has four outputs, which are labeled A, B, C, and D. Output A is the LSB and D is the MSB. This counter has a 0000_2 clear and a 1001_2, or 9_{10}, reset. A decade output count can be removed from the D output. When this output changes from its on state to zero, it indicates a count of 10_2. The 7490 is a modulo-10 counter.

In this activity, you will:

1. Become familiar with the operation of a BCD counter.

2. Construct a BCD counter and evaluate its operation.

Equipment and Materials

Multimeter

DC power supply—5 V

7490 binary-coded-decimal counter

SPDT switches (2)

LEDs—20 mA, 2 V (4)

Resistors—100 Ω, 1/4 W (4)

Pulser switch or a debounced R-S flip-flop [7400 IC (1), 1-kΩ, 1/4-W resistors (2), SPDT switch (1)]

Circuit construction board

Procedure

1. Connect the BCD counter of Figure 102-1.

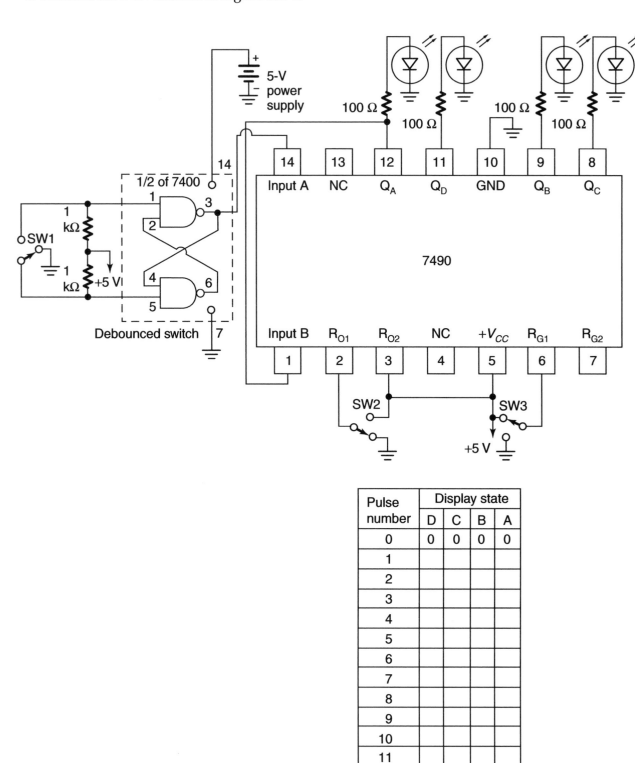

Figure 102-1. Circuit and truth table for a BCD counter.

Pulse number	Display state			
	D	C	B	A
0	0	0	0	0
1				
2				
3				
4				
5				
6				
7				
8				
9				
10				
11				
12				

2. The debounced switch, or pulser, is built with a 7400 quad NAND gate.

3. Turn on the power source and note the status of the LED display. If any one of the LEDs is on, it represents a random count. The count can be cleared by momentarily disconnecting pin 2, the R_{O1}, input from ground. SW2 controls this operation.

4. With the LEDs all off, change the debounced switch of the 7400 on and then off. Note the response of the display. This is generally referred to as *pulsing*. The 7490 changes states on the negative-going state change. The pulsing switch operation has two states: high and low. The state of the counter changes on the negative transition of the debounced switch.

5. Pulse the input of the counter several times, noting the response of the display.

6. Reset the display to 0000. Pulse the switch one time. Record the status of the LEDs in the table of Figure 102-1. An on LED indicates 1 and an off LED is 0. Pulse the counts indicated by the table recording the status of the LED display. The position of the LEDs is important in this display. LED Q_A is the LSB and LED Q_D is the MSB. Complete the table for the indicated number of pulses. Note that the last count that appears on the display before it resets to 0000. This is _____.

7. Pulse the counter so that the display shows 0011. Then clear the display with the R_{O1} switch. The clearing operation can be achieved on any count. Try some other count and clear the display. The letter designation R_O refers to "reset to 0."

8. Pulse the number 0101 into the counter. Momentarily open the SW3. How does this operation alter the display? _____

_____

Try several other numbers and reset the display with the R_{O1} switch. How does the display respond? _____

_____

9. The designation R_{G1} refers to "reset to 1." The 7490 has two R_G inputs. Either can be used independently or they can be used in combination. Resetting to 1 permits some math operations to be achieved automatically.

10. Turn off the power and remove the connecting wire between pins 1 and 12. Change the debounced switch from input A (pin 14) to input B (pin 1). Disconnect LED Q_A from pin 12.

11. Turn on the power. Reset the counter to 0000. Pulse the counter while observing the count on the display. LED Q_B is now the LSB and LED Q_D is the MSB. Indicate the counting sequence: ___ ___ ___ ___ ___ ___. This represents a modulo-___ counter.

12. Turn off the power. Remove the debounced switch from input B (pin 1) and connect it to input A (pin 14). Connect an LED to output Q_A as in the original circuit.

13. Turn on the power. Reset the counter to 0000.

14. Pulse the counter one time while observing the display. How does the display respond? ___

_____

Pulse the display several times while observing the display. Describe the response of the display. _____

_____

This represents what math function? _____

15. Turn off the power and disconnect the circuit. Return all materials.

Analysis

1. In order for a binary counter to be converted to a BCD counter, how many counts must it lose?

   _____

   _____

2. Why is BCD counting an important operation for a digital system? _____

   _____

   _____

Activity 103–Octal and Hexadecimal Counters

Name _____ Date _____ Score _____

Objectives

Binary-coded-octal (BCO) and binary-coded-hexadecimal (BCH) counters are similar in many respects to the BCD counter. A BCO counter displays a maximum count of seven (or 111_2) before it clears and displays all 0s.

When three separate BCO counters are ganged together, the maximum count will be $111,111,111._2$, or $7 \times 64 + 7 \times 8 + 7 \times 1$, or 511. Three BCH counters, by comparison, connected together can produce a maximum count of $1111,1111,1111._2$, or $15 \times 256 + 15 \times 16 + 15 \times 1$, or 4095. BCO and BCH counters permit the manipulation of very large numbers that would be difficult to achieve with binary alone.

In this activity, you will construct a simple BCO counter, and a BCH counter, with LEDs used to display the output. Two or more of these counters are then ganged together to achieve a greater counting capability. Through this activity you will be able to observe the actual counting of octal and hexadecimal numbers, thus becoming more familiar with these counting methods.

Equipment and Materials

DC power supply—0-5 V, 1 A

SN7493 counter IC

SE/NE555 timer/clock IC

Capacitors—0.01 µF, 100 Vdc; 4 µF, 25 Vdc

Resistors—470 Ω, 1/8 W (4); 200 kΩ, 1/8 W (2)

Light-emitting diodes (4)

SPST switch

IC circuit construction board

Procedure

1. Construct the binary-coded-octal (BCO) counter of Figure 103-1.

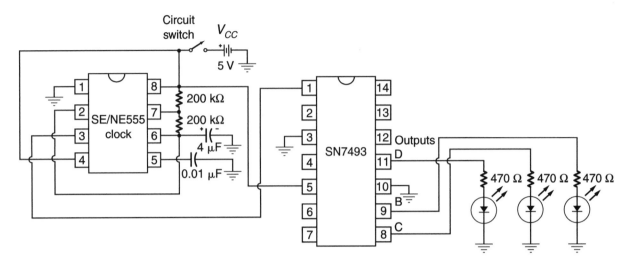

Figure 103-1. BCO counter and clock.

2. Before closing the circuit switch, turn on the power supply and adjust it to 5 Vdc.

3. Close the circuit switch, thus energizing the clock and the counter.

4. Explain the counting sequence displayed by the three LEDs. Label the LEDs A, B, and C from least significant digit to most significant digit, starting at the right and moving to the left. ____

___

5. Working with another lab group or with additional components, duplicate the counter part of the circuit shown in Figure 103-2 and connect it to the original counter as indicated.

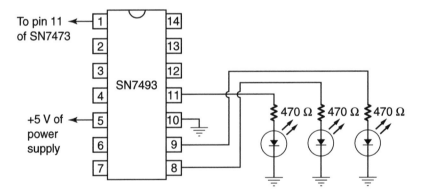

Figure 103-2.   Additional counter stage.

6. Close the circuit power switch and describe the counting action. _____

   _____

   The counter of Figure 103-1 will register units, while the one in Figure 103-2 will register 8s. The maximum count that can be displayed by this two-stage counter is _____.

7. Open the circuit switch and alter the construction of the counter to conform with the circuit of Figure 103-3. Only the modifications are shown; the remainder of the circuit is unchanged.

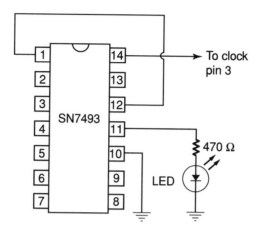

Figure 103-2. Additional counter stage.

8. Close the circuit switch and describe the counting action. _____

   _____

   The added LED should appear as the least significant digit and be located to the right of the original three. What is the maximum count produced by the LED display? _____

9. Working with another lab group or with additional components, duplicate the counter part of the circuit. Connect the clock input of the alternate counter to pin 11 of the counter of Figure 103-3. Use four LED readouts as in the modified circuit.

10. Close the circuit switch and describe the counting sequence. _____

    _____

    Consider the original counter as the units part and the additional counter as the 16s counter. The maximum count that can be achieved is _____.

11. If time permits, you may wish to connect a third counter to the two-stage counters. The maximum count that can be achieved by the three-stage counters is _____.

12. Open the circuit switch and disconnect the circuit. Return all parts.

Analysis

1. What is meant by the term *binary-coded-octal counter?* _____

_____

2. What is meant by the term *binary-coded-hexadecimal counter?* _____

_____

3. Why does the second stage of a BCD or BCH counter change states when the first stage changes from all 1s to 0s? _____

_____

_____

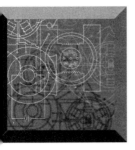

Activity 104–Digital Counting Systems

Name _____ Date _____ Score _____

Objectives

A digital counting system is the part of a digital system that is responsible for counting data and developing a readout display of the applied signal. Each new count applied to the system causes it to advance the previous count by one. Essentially the counter is the last part of a digital system. Functionally this part of the system is designed to change digital signals into information that will initiate a stage change of the readout. Binary signals are then decoded and used to produce a display of numbers.

There are four major sections of a counting system: signal source, counter, decoder-driver, and readout device. The counting signal for this system is generated by a clock circuit. In practice, this could come from a photoelectric detector, an electromechanical actuator, or some other physical transducer. The counter then responds to this signal by advancing once with each count. In this case, the counter has a binary input and BCD output. The BCD signal is then decoded and used to energize appropriate segments of the readout display.

The counting system used in this activity can only be used to count from 0 to 9. An output from the BCD circuit could, however, be used to initiate a 10s, 100s, 1000s, etc., count according to the demands of the system.

In this activity, you will build a one-digit counting system from discrete ICs. Each part of the system, starting with the readout, is built and tested. A similar counter built on a PC board is also tested and used to increase the count capability from 9 to 99, and possibly to 999. You will see how several counters are physically ganged together to increase counter size. The counting system principle is an extremely important concept today with the growing emphasis upon digital instruments.

Equipment and Materials

- DC power supply—5 V, 1 A
- SN7490 IC
- SN7447 IC
- MAN-1 seven-segment readout
- Resistors—470 Ω, 1/8 W (7)
- SPDT logic switches (4)
- SPST toggle switch
- Single-phase counter circuit board
- IC circuit construction board
- Clock circuit [SN/NE555 IC, 0.01-μF, 100-Vdc capacitor, 4-μF, 25-Vdc capacitor, 200-kΩ, 1/8-W resistors (2)]

Procedure

1. Connect the seven-segment LED readout of Figure 104-1.

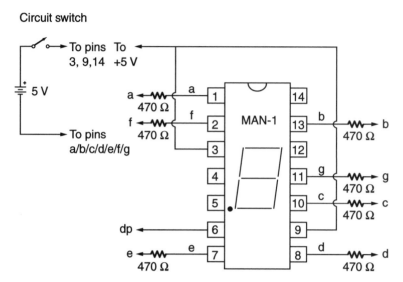

Figure 104-1. Seven-segment LED readout.

2. Close the circuit switch and test the negative side of the operation of each segment of the read-out by momentarily touching each to the 5-V supply. If each segment is lit up properly, proceed to the next step. If not, test the circuit wiring for a possible error.

3. Connect the SN7447 seven-segment decoder/driver circuit of Figure 104-2.

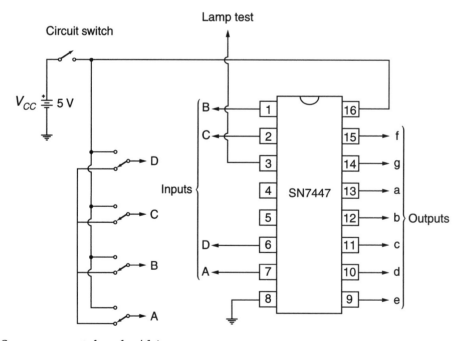

Figure 104-2. Seven-segment decoder/driver.

4. The lowercase letters at the right of the IC denote outputs that go to the readout accordingly. Connect the readout and SN7447 to the power supply.

5. Close the power supply circuit switch and momentarily connect the lamp test (pin 3) to ground. If the circuit is working properly, all seven segments will light up. Disconnect the lamp test wire from the circuit.

6. The A, B, C, and D logic switches are designed to control the numbers displayed on the readout. They are binary coded so that all 0s will produce a 0 display.

7. Test the decoder-driver by placing the appropriate binary code into the switches that will produce the numbers 0 through 9 on the display. Counts 10 through 15 will produce coded indications for the numbers 10 to 15. Turn off the circuit switch and remove the four logic switches attached to A, B, C, and D.

8. Construct the SN7490 BCD counter of Figure 104-3.

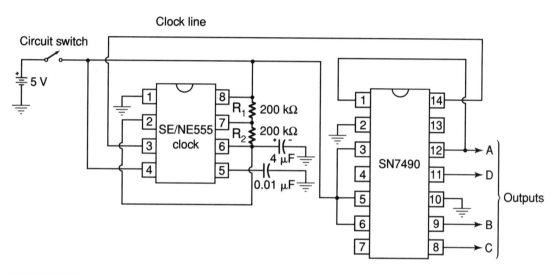

Figure 104-3. BCD counter circuit.

9. Connect the A-B-C-D outputs of the SN7490 to A-B-C-D inputs of the SN7447 where the logic switches were connected. Close the circuit switch. If the counting system is operating properly, the numbers 0 through 9 will be repeatedly displayed. If it does not display this counting procedure, check the clock circuit and the wiring of the SN7490. Open the circuit switch.

10. If additional counting systems are available, you may want to connect a second unit to the circuit board. The clock input of the second board should be connected to the D output of the SN7490. When both boards are energized, a 99 count can be achieved. The count can be increased to 999 by adding the third circuit board. The clock input of the third board should be driven by the D output of the SN7490 of the second board. The process can be repeated by ganging additional boards together.

11. Disconnect the counting system from everything but the clock. Connect the counting system circuit board to the positive and negative sides of the 5-V power supply. Connect the clock to the circuit board and turn on the power supply. If the circuit functions properly, the counter should display a repeat number count of 0–9.

12. Turn off the power supply and disconnect the circuit board and clock circuit. Return all parts.

Analysis

1. Describe the function of the four basic parts of a digital counting system. _____

   _____

   _____

   _____

2. Explain why the D output of an SN7490 can be used to initiate a count for the next stage of a counter. _____

   _____

   _____

Activity 105–Oscilloscope Measurement

Name _____ Date _____ Score _____

Objectives

The oscilloscope is one of the most important laboratory instruments in use today. It is essential for a technician to be able to use the oscilloscope competently. Oscilloscopes can be used to measure voltages, frequency, phase relationships, or to monitor waveform displays of various types.

In this activity, you will use the oscilloscope to measure the amplitude of an ac sine-wave voltage. The amplitude of an ac sine wave can be measured in three different ways: (1) peak-to-peak (p-p), (2) peak, and (3) root-mean-square (rms) or effective value. Peak-to-peak, as the term implies, is the measurement from the peak of the positive half of the waveform to the peak of the negative half, while peak is simply the measurement of either the positive or negative half. Root-mean-square voltage is the measurement of the amount of ac voltage that will do the same amount of work as a comparable dc voltage. The power-line voltage in the home, for example, is stated as an rms sine wave voltage, but the oscilloscope can measure voltages regardless of waveform type. This is a primary advantage of the oscilloscope.

Equipment and Materials

- Multimeter
- Oscilloscope
- AC power source
- Resistors—270 Ω, 470 Ω, 1000 Ω

Procedure

1. Turn on the power supply and adjust the voltage to 10 volts ac. Calculate the peak and p-p values of this voltage.

   Peak value = _____ volts ac.

   Peak-to-peak value = _____ volts ac.

2. Construct the circuit shown in Figure 105-1 and connect it to the 10-volt ac source. Measure the voltage across the following points with the meter. Calculate the peak and p-p values.

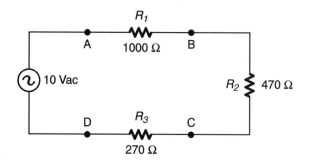

Figure 105-1. Test circuit for ac voltage measurements.

Points A to B: _____ volts rms; _____ volts peak; _____ volts p-p.

Points B to C: _____ volts rms; _____ volts peak; _____ volts p-p

Points C to D: _____ volts rms; _____ volts peak; _____ volts p-p

3. Prepare the oscilloscope for operation by adjusting the appropriate controls. Connect the ground lead to point D and the vertical-input lead to point A. Properly center the waveform and adjust the horizontal frequency and sync controls to produce two sine waves. Adjust the vertical controls so that each square or horizontal line of the graticule on the scope screen equals a definite p-p voltage. As previously determined mathematically, the p-p voltage of the 10-volt source is approximately 28.2 volts. Calibrate the scope so the sine wave has a specific number of squares or lines of amplitude. Once the scope has been calibrated, do not adjust the vertical controls again. (Note: Some scopes have internal calibration and the preceding is not required.)

4. With the scope calibrated, measure the p-p voltage across the following points and record the values:

Points A to B = _____ volts p-p.

Points B to C = _____ volts p-p.

Points C to D = _____ volts p-p.

Analysis

1. In order to measure p-p voltage, the scope must first be calibrated, then it can be used to make any p-p measurements. How does changing the vertical attenuation or volts/division control affect the calibration? _____

_____

_____

_____

2. Discuss the basic operation of the oscilloscope. _____

_____

_____

_____

3. How can an oscilloscope be used to measure frequency? _____

_____

_____

_____

4. Explain the procedure for calibrating an oscilloscope to measure p-p ac voltage. _____

_____

_____

_____

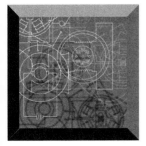

Activity 106–Chart Recorders

Name _____ Date _____ Score _____

Objectives

Chart recorders are used to monitor many types of electrical and physical quantities over a period of time. The advantage of a chart recorder is that a visible record of the monitored quantity can be kept for future use. There are many different types of chart recorders in use today.

In this activity, you will be able to study the construction and operation of a chart recorder. The recorder should be connected to some operational system so that you can observe its operation.

Equipment and Materials

- Chart recorder (any type available)

Procedure

1. Obtain a chart recording instrument and operating manual (if available). Record the manufacturer and model number.

   Manufacturer _____.

   Model Number _____.

2. What quantity or quantities does this instrument measure? _____

   _____

   _____

3. What is the range or scale of this instrument? _____

   _____

   _____

4. What method is used by this instrument to make a permanent record of the quantity measured?

   _____

   _____

   _____

5. What type of drive mechanism is used on this instrument? _____

   _____

6. What type of chart is used by this instrument? _____

   _____

7. Describe any special features this instrument has._____

_____

_____

8. Prepare the chart recorder for operation. Connect the recorder to some circuit or system in which you can observe its operation over its entire range of measurement.

9. Have your instructor check the instrument for operation.

   Instructor's Approval: _____

Analysis

1. What are some specific applications where chart recorders might be used? _____

_____

_____

2. What are some types of drive mechanisms used for chart recorders? _____

_____

_____

3. What are some methods used by chart recorders for recording the measured quantity? _____

_____

_____

Activity 107–Single-Phase Power Measurement

Name _____ Date _____ Score _____

Objectives

In direct-current circuits, power (watts) is always equal to the product of voltage and current. This relationship is also true for resistive ac circuits. However, when reactive ac circuits that contain either inductance or capacitance are encountered, the power converted in the circuit is no longer equal to the product of voltage and current. It is necessary to use a wattmeter to measure the true power (W) of these types of ac circuits. The product of voltage and current is referred to as the apparent power (VA) of a circuit. The power delivered from the source, but not converted to another form of energy by the load, is called reactive power (VAR). These three components form the power triangle relationship.

In this activity, you will use a single-phase wattmeter to measure the true power of a motor. You will be able to measure the voltage, current, and power associated with the operation of a single-phase ac motor.

Equipment and Measurement

- Multimeter
- AC current meter
- Wattmeter—single-phase (2000 W or higher)
- AC power source—120 V, single-phase
- Single-phase ac induction motor

Safety

Be sure to wear eye protection while electric motors are in operation. Also, be very careful when working with high voltages.

Procedure

1. Obtain a single-phase ac motor of any type and a single-phase wattmeter.

2. Check the nameplate data on the motor. Multiply the rated voltage of the motor by its rated current. Record the values.

    Rated voltage = _____ volts ac.

    Rated current = _____ amperes ac.

3. Multiply the voltage and current values to obtain a full-load apparent-power rating.

   Apparent power = _____ VA.

4. Make sure that the wattmeter will measure this amount of power.

5. Connect the wattmeter to the motor as shown in Figure 107-1.

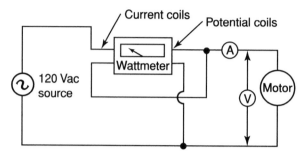

Figure 107. Test circuit for single-phase power measurement.

6. Measure and record the ac voltage (V) applied to the motor.

   Voltage = _____ volts ac.

7. Apply power to the motor. Observe the reading on the wattmeter and current meter. Record the maximum power and current during the starting period.

   Starting power = _____ watts.

   Starting current = _____ amperes ac.

8. Measure and record the power and current while the motor is running.

   Power = _____ watts.

   Current = _____ amperes ac.

9. Apply a load to the shaft of the motor while observing the meters. Record the maximum power and current.

   Maximum power = _____ watts.

   Maximum current = _____ amperes ac.

Analysis

1. How is a single-phase wattmeter constructed? _____

   _____

   _____

2. From the data of Step 7, calculate:

   a. Apparent power = volts × amperes = _____ VA.

   b. Power factor = $\dfrac{\text{true power}}{\text{apparent power}}$ = _____.

   c. Reactive power = $\sqrt{VA^2 - W^2}$ = _____ VAR.

3. From the data of Step 8, calculate:

   a. Apparent power = _____ VA.

   b. Power factor = _____.

   c. Reactive power = _____ VAR.

4. From the data of Step 9, calculate:

   a. Apparent power = _____ VA.

   b. Power factor = _____.

   c. Reactive power = _____ VAR.

5. As the load is increased on the single-phase ac motor, what happens to:

   a. Apparent power? _____

   _____

   b. True power? _____

   _____

   c. Power factor? _____

   _____

   d. Reactive power? _____

   _____

Activity 108–Three-Phase Power Measurement

Name _____ Date _____ Score _____

Objectives

Two single-phase wattmeters can be used to measure the power converted by a three-phase load. When the meters are properly connected, the algebraic sum of the two readings will equal the total three-phase power converted by the load. However, if the power factor of the load device is less than 50%, one wattmeter will deflect in the wrong direction. In this case, the current-coil connections on the meter should be reversed so that the meter will deflect properly. Then, the lower meter reading will be subtracted from the higher one.

In this activity, you will use two single-phase wattmeters to measure the power converted by a three-phase motor.

Equipment and Materials

* Multimeter

* AC current meter

* AC power source—three-phase

* Wattmeters—single-phase (2)

* Three-phase ac induction motor

Procedure

1. Obtain a 208-volt or a 240-volt three-phase ac motor and connect two single-phase wattmeters to it as shown in Figure 108-1.

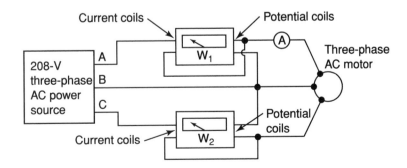

Figure 108-1. Test circuit for three-phase power measurement.

2. Measure and record the ac voltage applied to the motor.

Voltage = _____ volts ac.

3. Apply power to the motor. Observe the wattmeter and ammeter readings. Record the maximum value of each when the motor starts.

Starting power (wattmeter 1) = _____ watts.

Starting power (wattmeter 2) = _____ watts.

Current = _____ amperes ac.

4. Measure and record the same values while the motor is running.

Power (wattmeter 1) = _____ watts.

Power (wattmeter 2) = _____ watts.

Current = _____ amperes ac.

5. Apply a load to the motor shaft while observing the meters. Record the maximum power and current.

Maximum power (wattmeter 1) = _____ watts.

Maximum power (wattmeter 2) = _____ watts.

Maximum current = _____ amperes ac.

Analysis

1. What are some ways in which three-phase power can be measured? _____

_____

_____

2. From the data of Step 3, calculate:

a. True power = $W_1 + W_2$ = _____ watts.

b. Apparent power = $V \times A \times 1.73$ = _____ VA.

c. Power factor = $\dfrac{\text{true power}}{\text{apparent power}}$ = _____.

d. Reactive power = $\sqrt{VA^2 - W^2}$ = _____ VAR.

3. As load is increased on a three-phase motor, what happens to:

a. Apparent power. _____

b. True power. _____

c. Power factor. _____

d. Reactive power. _____

4. From the data of Step 4, calculate:

   a. True power = _____ watts.

   b. Apparent power = _____ VA.

   c. Power factor = _____.

   d. Reactive power = _____ VAR.

5. From the data of Step 5, calculate:

   a. True power = _____ watts.

   b. Apparent power = _____ VA.

   c. Power factor = _____.

   d. Reactive power = _____ VAR.

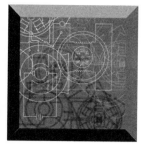

Activity 109–Watthour Meter Familiarization

Name _____ Date _____ Score _____

Objectives

Watthour meters are used to monitor the amount of electrical power used over a period of time. Any industrial plant or residence that uses electrical power has a watthour meter to measure the power used at that particular location.

In this activity, you will be able to observe the operation of a watthour meter. You can connect a load to the meter (such as a motor) to cause it to function. If a meter is not available in the lab, you can observe an operational watthour meter either near the laboratory or at home.

Equipment and Materials

• Watthour meter—single-phase

• Motor—single-phase ac (or other load device)

Procedure

1. Connect a watthour meter to the load as shown in Figure 109-1.

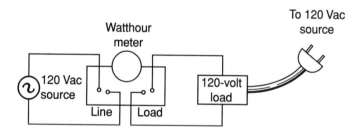

Figure 109-1. Circuit for observing the operation of a watthour meter.

2. Apply power to the load and observe the operation of the watthour meter. Note the speed of rotation of the aluminum disk.

   Speed = _____ r/min (approximately).

3. Increase the load while observing the speed of the aluminum disk.

4. Have your instructor check your meter for operation.

   Instructor's Approval: _____

Analysis

1. Discuss the operational principle of a watthour meter.

_____

_____

2. In Step 3, what happened to the speed of the aluminum disk? Why?

_____

_____

Activity 109—Watthour Meter Familiarization

Activity 110–Dynamometer Analysis of Electric Motors

Name _____ Date _____ Score _____

Objectives

A very thorough analysis of the operation of any motor under varying load conditions can be made by using a dynamometer. The dynamometer unit should be used to analyze and compare the characteristics of many types of fractional horsepower motors. Most commonly, 1/4-, 1/3-, or 1/2-horsepower motors are used on dynamometer units.

In this activity, you should analyze as many motors as possible. There is no particular order in which they should be done. When you complete an analysis, you should submit to the instructor the following:

1. Motor nameplate data.

2. Motor analysis data.

3. Motor data graph with the following performance curves.

   a. Speed vs torque.

   b. Speed vs horsepower.

   c. Efficiency vs torque.

4. A written analysis in two parts:

   a. General information about the operational characteristics of this type of motor.

   b. Reference to the data obtained to discuss the characteristics of the motor. (Discuss each characteristic curve on the graph.)

Equipment and Materials

- Multimeter

- Dynamometer unit

- Wattmeter

- AC ammeter

- Electric motor(s)

- Graph paper (one sheet for each motor)

- Calculator

Procedure

1. Select a motor for analysis and record the nameplate data of the motor in Figure 110-1.

```
Motor type _____

Manufacturing Co. _____

Identification number _____

Model number _____

Frame type _____

Number of phases (ac) _____

Horsepower _____

Cycles (ac) _____

Speed (r/min) _____

Voltage rating _____

Current rating (amperes)_____

Thermal protection _____

Temperature rating–°C _____

Time rating _____

Other information _____
```

Figure 110-1. Motor nameplate data.

2. Secure the motor onto the dynamometer unit. Make sure the shaft couplings are of the proper type for the motor you are analyzing.

3. Connect the motor to the current and power meters as shown in Figure 110-2. Be sure to connect the wattmeter properly.

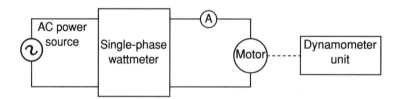

Figure 110-2. Connection diagram for dynamometer analysis of electric motors.

4. Adjust the load applied to the motor by the dynamometer through as many increments of torque as possible. Record the current and power for each value of torque in Figure 110-3.

5. Using the data recorded in Figure 110-3, and the rated voltage of the motor, calculate the horsepower, efficiency, and power factor for each value of torque. Record these calculated values in the appropriate columns of Figure 110-3.

Activity 110—Dynamometer Analysis of Electric Motors

Torque (ft · lb)	Speed (r/min)	True power (watts)	Line current (amperes)	Horsepower*	% Efficiency†	Power factor‡
1						
5						
10						
15						
20						
25						
etc.						

* Horsepower = $\dfrac{\text{torque (ft · lb) x speed (r/min)}}{5250}$

† % Efficiency = $\dfrac{\text{horsepower x 746}}{\text{true power (watts)}}$

‡ Power factor = $\dfrac{\text{true power (watts)}}{\text{apparent power (voltamperes)}}$

Figure 110-3. Motor analysis data.

6. On a sheet of graph paper plot the performance curves for the motor data graph by using the values of your recorded data. Refer to the sample graph of Figure 110-4 for selection of the proper horizontal and vertical axes.

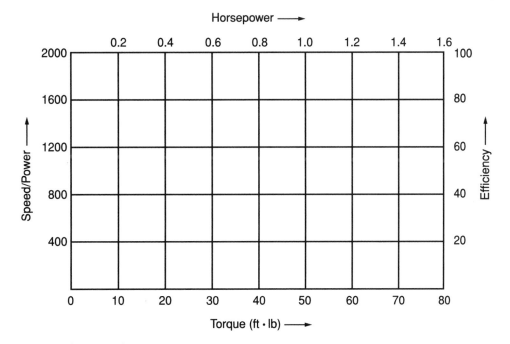

Figure 110-4. Motor data graph.

7. Prepare your data and analysis to be submitted for evaluation. Try to do a neat and accurate job on your data and graph, and a detailed job on the analysis.

8. Perform the same procedure on as many motors as possible. It may be interesting to compare the curves obtained from several different types of motors.

Analysis

1. Discuss the use of a dynamometer to analyze an electric motor. _____

_____

_____

_____

2. Complete all of the information listed in the Objectives of this activity.

Appendix-Simulation Software

Several types of software are available to simulate actual control circuits and make tests and measurements. This software may be used to supplement the activities of this manual and to enhance learning about electrical control systems.

Recommended software to supplement *Electrical Motor Control Systems* and this accompanying Laboratory Manual are listed below.

1. **MotorCon Tutor**—distributed by:

   ECHOSCAN Inc.

   1402 Pine Ave.

   PMB 5263

   Niagara Falls, NY 14301

   (905) 371-1645

This software is designed to supplement motor control. The following units are included:

- Introduction to Motor Controls
- Two-Wire Controls
- Three-Wire Controls
- Separate Controls
- Hands-Off Automatic Controls
- Sequence Controls
- Reversing Controls
- Jogging Controls
- Timing Relays
- Starting Methods
- Deceleration Methods
- Programmable Logic Controllers

2. **Laboratory Manual for Robotics Technology**—distributed by:

   Goodheart-Willcox Publisher

   18604 West Creek Drive

   Tinley Park, Illinois 60477

   (800) 323-0440

This manual contains IBM AML Programming Software, which is modified for educational use. Simulation programs can be performed without actually purchasing an industrial robotic workstation. This is commonly done in industry.

3. **Electronics Workbench**—distributed by:

   Interactive Image Technologies, Ltd.

   111 Peter St., Suite 801

   Toronto, Ontario, Canada

   M5V 2H1

   (800) 263-5552

This is powerful software that enables you to easily build and simulate many types of analog and digital circuits. On-screen instruments are the same as those you would use in the laboratory. It is easy to use and great for learning about electrical circuits and measurements.

Appendix-Simulation Software